Industrial Disasters and Environmental Policy

Industrial Disasters and Environmental Policy

Stories of Villains, Heroes, and the Rest of Us

Denise L. Scheberle

Routledge
Taylor & Francis Group

NEW YORK AND LONDON

First published 2018 by Westview Press

Published 2018 by Routledge
711 Third Avenue, New York, NY 10017, USA
2 Park Square, Milton Park, Abingdon, Oxon OX14 4RN

Routledge is an imprint of the Taylor & Francis Group, an informa business

Copyright © 2018 Taylor and Francis

Print book interior design by Timm Bryson, em em design, LLC

Library of Congress Cataloging-in-Publication Data has been applied for.

ISBN: 978-1-138-58916-2 (hbk)
ISBN: 978-0-8133-4725-7 (pbk)
ISBN: 978-0-429-46880-3 (ebk)

LSC-C

This book is dedicated to my family, my colleagues, and my students—so many of whom are environmental heroes.

Contents

List of Illustrations

List of Abbreviations

Agency for Toxic Substances and Disease Registry (ATSDR)
Bureau of Land Management (BLM)
Bureau of Ocean Energy Management (BOEM)
Bureau of Ocean Energy Management, Regulation,
    and Enforcement (BOEMRE)
Bureau of Safety and Environmental Enforcement (BSEE)
Centers for Disease Control and Prevention (CDC)
Central Bureau of Investigation (CBI)
Chemical Manufacturers' Association (CMA)
Clean Air Act (CAA) of 1970 (CAA)
Comprehensive Environmental Response, Compensation,
    and Liability Act of 1980 (CERCLA, the Superfund law)
Conferences of the Parties (COP)
Council on Environmental Quality (CEQ)
dichloro-diphenyl-trichloroethane (DDT)
Democratic National Committee (DNC)
Emergency Planning and Community Right-to-Know Act
    of 1986 (EPCRA)
Environmental Impact Statement (EIS)
Goddard Institute for Space Studies (GISS)
Governor's Independent Investigation Panel (GIIP)
Highly Immersive Visualization Environment (HIVE)
high-reliability organization (HRO)
Intergovernmental Panel on Climate Change (IPCC)
International Program on Chemical Safety (IPCS)
Libby amphibole asbestos (LA)
Lifestyles Of Health And Sustainability (LOHAS) consumers
local emergency planning committees (LEPCs)
material safety data sheet (MSDS)

methyl isocyanate gas (MIC)
Mine Improvement and New Emergency Response (MINER) Act of 2006
Minerals Management Service (MMS)
Mine Safety and Health Administration (MSHA)
National Aeronautics and Space Administration (NASA)
National Ambient Air Quality Standards (NAAQSs)
National Environmental Policy Act (NEPA)
National Emissions Standards for Hazardous Air Pollutants (NESHAP)
National Institute of Occupational Safety and Health (NIOSH)
National Oceanic and Atmospheric Administration (NOAA)
National Priorities List (NPL)
natural resource damage assessment (NRDA)
nongovernmental organizations (NGOs)
Occupational Safety and Health Administration (OSHA)
Office of Natural Resources Revenue (ONRR)
remotely operated vehicles (ROVs)
Resource Conservation and Recovery Act (RCRA)
Resources and Ecosystems Sustainability, Tourist Opportunity,
     and Revived Economies of the Gulf States Act of 2011
     (the RESTORE Act)
Robert C. Byrd Mine Safety Protection Act (MINER) of 2010
self-contained self-rescue device (SCSR)
terawatt-hours (TW-h)
Toxics Release Inventory (TRI)
Triple Bottom Line (TBL)
Union Carbide Corporation (UCC)
Union Carbide India Limited (UCIL)
United Mine Workers of America (UMWA)
Upper Big Branch (UBB) mine
US Chemical Safety and Hazard Investigation Board (CSB)
US Coast Guard (USCG)
US Department of Agriculture (USDA)
US Energy Information Administration (EIA)
US Environmental Protection Agency (EPA)
US Office of Surface Mining Reclamation and Enforcement (OSMRE)
West Virginia Department of Environmental Quality (WVDEQ)
West Virginia Office of Miners' Health Safety and Training (WVMHST)
World Wildlife Fund International (WWF)

Preface

The idea for this book came on a spring day in 2010. I had just described to my students the tragedy of the Upper Big Branch mine explosion that had recently killed twenty-nine miners, on April 5. Massey Energy, headed by Don Blankenship, had a reputation for running coal at all costs, even if it meant harming the environment and endangering miners working more than a thousand feet underground. This was the worst mining accident in West Virginia since 1968.

About two weeks later, on April 20, 2010, what would come to be known as the worst environmental disaster in US history occurred. BP had attempted to seal the Macondo well a mile beneath the Gulf of Mexico, but failed. As a result, high-pressure gas shot through the riser to the *Deepwater Horizon* oil rig, where it ignited and soon engulfed the massive rig. The rig sank two days later, and eleven missing workers were never found. But that was not the extent of the disaster. As many of us watched the live television feed, the oil leaking from the failed well would eventually spew over 3.19 million barrels of oil into the Gulf of Mexico.

Investigations began for both accidents, and it soon became apparent that these were more than just accidents one might expect when conducting dangerous operations such as mining for coal deep underground or drilling for oil a mile beneath the sea. Instead, these two accidents were the result of misconduct by individuals within Massey and BP who put lives at risk in the pursuit of profitability. Criminal charges were soon brought against these companies, and my students and I began to wonder if commonalities existed between these two tragedies.

Fast-forward just another few weeks: on June 7, 2010, a court in India finally sentenced eight people to two years in jail for their role in another explosion. The 1984 explosion of deadly methyl isocyanate gas in Bhopal, India, took thousands of lives and did lasting environmental damage around the Union Carbide pesticide manufacturing facility in that city. The court found that these plant employees were guilty of "death by negligence." And, again, I wondered what connections might be drawn to the Upper Big Branch mine and BP oil spill disasters.

Then another recent event came to mind: the declaration by the US Environmental Protection Agency (EPA) on June 18, 2009, that Libby, Montana, represented a public health emergency. This declaration was the first time the EPA had made such a determination under the authority of the 1980 Superfund law, and it had triggered a substantial federal cleanup of the town. The reason for this massive cleanup was a vermiculite mine owned by W. R. Grace. Vermiculite is laced with asbestos, and company executives and mine managers knew it. Mine operations had imposed a dear cost on the families of the town, as hundreds died as a consequence of exposure to vermiculite and many hundreds more have suffered from asbestos-related illnesses.

Four stories, each unique, but also with a seemingly similar thread of disregard for human and environmental harm. I wanted to learn more about what lay behind each apparently villainous story.

At the same time, these stories were almost overwhelmingly sad; told within the scope of environmental policy, they didn't tell the whole story of the individuals and organizations involved. In teaching environmental policy, management, and law courses over several decades, I've learned that not all companies are bad actors. Increasingly, many companies are adopting a triple-bottom-line approach (valuing the environment and people as well as profits) and moving to more sustainable practices. And some organizations with high-risk operations have been resilient and free of tragic mishaps. I wanted to learn the stories of these four tragedies, with an eye toward the organizational and individual behavior animating them.

I also wanted to see what these stories might suggest about long-standing concepts in public policy in general, and environmental policy specifically. For example, what was the policy response after these events? What happened to laws, to regulatory agencies, and to the companies? I wanted to dive more deeply into the scholarship of "normal" accidents, organizational deviance, and high-reliability organizations. As a student of public policy, I also wanted to know more about government's response to these tragedies: to what extent did these tragedies bring new policies to the forefront of the congressional and legislative agendas or change regulatory approaches to these various industrial activities?

Moreover, I wanted to know if we as a society can learn from industrial disasters so that they might become ever rarer, and what public organizations and individuals—"the rest of us"—can do to help prevent them. These four stories appeared in the news in some fashion at about the same time, and all of them pointed to less than stellar organizational performance on the part of the companies and the regulators tasked with overseeing them. Minimizing incidents like these must be a concerted effort.

Beyond these disasters, I wanted to learn more about a few individuals who represented what I referred to in my classes as "environmental spark plugs." These were people who stood up for the environment in a way that exhibited courage, integrity, creativity, and environmental values. I chose one senator (Gaylord Nelson), one EPA administrator (William Ruckelshaus), and one grassroots environmental activist (Judy Bonds). Gaylord Nelson, the father of Earth Day, showed a creative spark and the genius of an idea that captured the lasting attention of a country. William Ruckelshaus, the first and fifth administrator of the EPA, represented for me an individual with integrity and common sense. Judy Bonds, winner of the 2003 Goldman Prize honoring the achievement of a grassroots environmental activist, had boundless energy and great courage. All three sought to engage the rest of us in environmental protection efforts.

I reasoned that if we could understand what motivated these individuals, we all might one day spark environmental protection in our communities, cities, states, and countries. So the book ends on a happier note, with a vision of what is possible and what can be achieved even in the thorniest of environmental challenges—climate change—when we take up the mantle of an everyday environmental hero.

I did not do this alone. I am indebted to the reviewers for their helpful feedback, including: Stefanie Chambers (Trinity College); Robert C. Turner (Skidmore College); Lada V. Kochtcheeva (North Carolina State University); Michelle Pautz (University of Dayton); Rebecca Bromley-Trujillo (University of Kentucky); Laurel E. Phoenix (University of Wisconsin–Green Bay); Jessica Weinkle (University of North Carolina–Wilmington); Sara Rinfret (University of Montana–Missoula); and others who wished to remain anonymous. I am especially indebted to Ada Fung, my editor, who encouraged me to think more deeply and holistically in organizing and writing this book. Thanks, too, to Amber Morris, senior project editor, and Cindy Buck, copy editor, from Westview Press, as well as Anna Dolan, production editor, Kay Mariea, proofreader, and the staff at Taylor and Francis who carefully shepherded the book's final stages. I thank my husband, Steve, who graciously left me to the keyboard and stood by me as

I went through the highs and lows of writing a book, my daughter, Jenni, and grandsons, Trystin and Owen, for encouraging me. Finally, thanks to my students, who have touched my heart and time and again showed me how environmental spark plugs should act. They give me hope that the future is in good hands.

CHAPTER 1

Telling Stories

VILLAINS AND HEROES IN
ENVIRONMENTAL POLICY

Once upon a time, in faraway lands as well as in our own communities, heroes and villains and the rest of us used, abused, and protected the environment. In big or small ways, each had a story to tell and each made an impact. As each story played out, some wondered if heroes would rise to the occasion, avoiding the kryptonite of disinterest to protect the environment and human life. Others thought about getting involved, but wondered what to do. Some stories ended badly. Great damage was done. Sometimes with villainous intent and sometimes simply with little regard for the consequences of their actions, people left in their wake spoiled landscapes, public health risks, endangered species, and even death. It was up to the rest of us to rebuild and recover, if we could.

Other environmental tales have ended with hope, having come to a more sustainable, environmentally sensitive conclusion. What makes the difference? It is persistence, integrity, wisdom, and courage that help shape happier endings to environmental stories.

The environmental stories that influence our environmental laws, regulations, and policies often have all the dramatic elements that make for a good tale. As with any good story, we are captivated by the high drama, the suspense, the danger—but most of all, by the cast of characters. Some of these stories are well known—who among us will soon forget the BP oil spill disaster in the Gulf of Mexico? Other, less familiar stories involved small communities and ordinary people doing extraordinary things. All of these stories, however, help illustrate how our environmental policies have come to be, how they protect public health and the environment, or why they have failed to work as intended.

This book takes us on a journey. It tells stories of actions taken by organizations, government agencies, and individuals. By no means do these represent the only such stories, and determining whether these are tales of villainous or heroic acts is ultimately up to the reader. But there are remarkable elements of these fascinating tales that ensure that they will be remembered for decades to come. By examining high-profile incidents, limning portraits of individuals and groups working to protect the environment, and detailing what we can do in our own part of the world, the book presents a way of seeing how environmental politics, policies, and laws work and how we can work within them to tell new stories.

In its consideration of extraordinarily tragic events, this book also focuses on extraordinarily courageous actions taken by environmental heroes (Chapter 6) and "the rest of us" (Chapter 7) for a simple reason: ultimately, earth's various environmental stories have to end with us. Environmental laws and regulations can go only so far in protecting the environment and people. Laws and regulations are the engines that drive environmental protection, but citizens are the spark plugs. We are the ones who decide whether or not to participate, whether to observe and support the actions of federal and state agencies, and, ultimately, whether to act to enforce the law or to lobby for additional protections. We can encourage companies, too, to promote sustainable behavior and even go above and beyond regulatory requirements to help protect the environment. By being part of any environmental story, we might just influence decision-makers to heed the warning signs of an industrial disaster in the making. We can join in or stand by and do nothing, watching the story unfold.

The heroes are among us, and the villains are both lurking about in the shadows and standing in full view. Will we join the heroes, or become heroes ourselves? Will we put up with the villains? In the end, it's really up to us. One way to begin is to tell the stories of four high-profile disasters (detailed in Chapters 2 to 5) that exacted a heavy toll in human life and ongoing injuries to the environment. Learning the lessons from these stories will potentially help us avoid similar industrial disasters in the future.

UNDERSTANDING ORGANIZATIONS AND WHY INDUSTRIAL DISASTERS HAPPEN

Industrial disasters are distinguished from natural disasters, such as hurricanes, wildfires, and floods, in that they are caused by organizations engaged in high-risk activities. These human-made disasters are of such a magnitude that they prompt major disruptions in the organization itself and sometimes spawn new policies, regulations, laws, and even agencies to govern that industry. To put it simply, these disasters are of sufficient size and scope to force changes in the way we view not only the organizations involved but the entire industry and the way it operates.

Astute readers may wonder why we should consider past accidents at all. Accidents happen, one might argue, and tragedies occur and always will, especially in our high-tech, fossil fuel– and chemical-driven society. Could it not be the case that some industrial disasters capture our attention because they are simply larger events than others? The answer to this is: sometimes . . . and sometimes not.

Distinguishing Normal Accidents from Avoidable Accidents

Scholars have long attempted to explain why organizations act in ways that discount safety or environmental issues. They attempt to draw distinctions between "simple" mistakes, on the one hand, and misconduct or criminal actions that give rise to the "dark side" of organizations, on the other.[1] Whether or not an event represents behavior that could be characterized as "dark" or reprehensible depends in part on how the issues are interpreted and defined. Making sense of these disasters has long captured the attention of scholars in the fields of public policy, political science, psychology, business, sociology, criminology, and organizational behavior, among others.

Charles Perrow argues that complex enterprises engaged in high-risk activities will always have catastrophic potential.[2] Accidents in these complicated systems should be expected and are therefore "normal." According to "normal accident" theory, two organizational characteristics, "interactive complexity" and "tight coupling," make these systems susceptible to accidents. Normal accidents may also be called "system accidents," where the system includes not only the equipment and other components but also the humans who operate them.

Any system, by definition, has many parts, any one of which may fail. In a simple system, single malfunctions, or "discrete failures," may be spotted and corrected. However, a complicated system may allow two or more discrete failures to interact in unexpected ways, thus creating what Perrow defines as "interactive complexity." In turn, these unexpected interactions can affect supposedly redundant or backup systems, creating a series of malfunctions that may lead to catastrophe in a blink of an eye. Perrow focused on the 1979 accident at the Three Mile Island nuclear plant in Pennsylvania; later, his "normal accidents" framework would be used to examine other incidents, such as the 1986 *Challenger* disaster, when the space shuttle broke apart shortly after launch, leading to the deaths of its crew members. We can assume that a sufficiently complex system, like a nuclear power plant or a space shuttle, is susceptible to normal accidents because it can be expected to have many such unanticipated interaction failures. However, normal accidents in these complex systems have been minimized by vigilant staff in high-reliability organizations, described later in this chapter.

Normal accidents are also triggered by tightly coupled system components. "Tight coupling" exists when system components are linked closely in time or space. If the system allows sufficient time after the discrete failure occurs, operators of the system are able to respond. In contrast, tightly coupled systems create a rapid chain of events, so that components have major impacts on each other in a short time frame. Because of these tight linkages, system operators have almost no time to react. Tight coupling raises the odds that the responses of decision-makers trying to correct the failure will be wrong, since they do not correctly understand the true nature of the problem. Because system components are interacting in unexpected ways, the problem is incomprehensible, at least for a short period of time. As a result, a cascade of decisions may amplify the tragedy, and each single decision may have a deleterious effect on the outcome. Sadly, the failure that initiates a catastrophic event often seems, taken by itself, quite trivial. Because of the system's complexity and tight coupling, however, events quickly surge out of control, creating a cataclysmic outcome.

By contrast, some accidents in complex systems are not "normal." Accidents result because of an organization's poor leadership, misdirected values, and a culture of complacency about safety and environmental regulations. Organization executives, driven by an all-encompassing desire to maximize profits or cut costs, seek to reach a goal regardless of the consequences.

Unintended Consequences, Organizational Culture, and Power

Why do organizations depart from their own goals, act in unethical ways that harm the public, or engage in criminal conduct? Answers to these questions can be found in research focused on unintended consequences, organizational culture, and the power and intent of the individuals responsible for an organization.

Let's begin by looking at unintended consequences. BP, Union Carbide, Massey Energy, and W. R. Grace, the companies whose stories are told in Chapters 2 to 5, undoubtedly never wanted the environmental tragedies for which they were responsible to happen. Those disasters were likely to have been the unintended consequence of poor decision-making based on overriding values, like maximizing profitability. Robert Merton was one of the first scholars to point to unanticipated consequences of individual conduct in organizations.[3] He observed that any purposive action inevitably generates unintended consequences, which may be positive or negative. He theorized that certain conditions make bad outcomes more likely and often will even exacerbate negative unintended consequences. These conditions include: failure to fully understand the problem; ignorance; attention to satisfying immediate interests rather than long-term goals; and any other unethical values that organizational actors might bring to decision-making.

Merton also suggested that a common fallacy is the "too-ready assumption that actions which have in the past led to the desired outcome will continue to do so."[4] That is, individuals commit errors in judgment because they fail to recognize that what has been successful in certain circumstances may not work under future conditions that are different. Error often involves either neglect or failure to thoroughly examine the situation, perhaps owing to what Merton called "pathological obsession": a determined refusal to consider all aspects of a problem. Merton also felt that emotional involvement could distort our construction of a situation and the probable consequences. Perhaps most vexing is the influence of what Merton referred to as the "immediacy of interest" on basic values. He suggested that we may go against fundamental values and fail to consider further consequences because of our intense interest in satisfying immediate desires.

Over sixty years later, Diane Vaughan arrived at similar conclusions about values, ignorance, and errors due to complacency. She has sought to explain why organizations go over to the "dark side"—why they behave in ways that deviate from their formal organizational goals and from standards and expectations for behavior.[5] She notes that people within an organization can become accustomed to a deviant behavior (such as ignoring safety rules) to the point where they don't consider it deviant, a phenomenon she calls "normalization of deviance." Then, as people inside an organization continue to regularly depart from accepted behavior, they grow more comfortable breaking the rules or relaxing safety standards.

If left unchecked long enough, organizational deviance results in un-corrected mistakes, misconduct, and, sometimes, even industrial disaster. Misconduct suggests intent: individuals or groups within an organization, acting in their organizational roles, intentionally violate internal rules, laws, or administrative regulations in pursuit of organization goals. They go against established procedures. The question is: why?

Part of the answer can be found in the culture of an organization. Or-ganizational culture creates the "rules of the road" for employees inside an organization to follow. Culture helps employees understand whether the or-ganization is sincerely interested in complying with regulations, protecting the environment, and putting the health and safety of its workers first—or not. Culture can also create that dark space into which normalization of deviance, or routine nonconformity, is accepted and even encouraged.[6] For example, it becomes expected that employees will support the organiza-tion's cost-cutting measures, even if doing so reduces attention to safety procedures.

In a famous example, Vaughan and others have argued that the 1986 *Challenger* disaster was caused by both a technical problem (the failure of the O-rings to seat properly in low temperatures) and an organizational cul-ture problem. Engineers and consultants at the National Aeronautics and Space Administration (NASA) were made to feel that their voicing of con-cerns would not be tolerated in an atmosphere where political pressure had made getting the space shuttle launched the highest priority and additional delays would not be welcome. As a consequence, they kept quiet about their concerns.[7] The Presidential Commission on the Space Shuttle *Challenger* Accident, known as the Rogers Commission, focused on NASA manage-ment and on what the commission would later reveal was an organizational culture that had gradually begun to accept escalating risk and whose safety program had become "largely silent and ineffective."[8]

This dysfunctional culture at NASA persisted for another seventeen years, until the space shuttle *Columbia* accident on February 1, 2003, that killed the seven-member crew. The *Columbia* Accident Investigation Board's independent investigation once again found both technical and or-ganizational causes of the accident. The board's report notes that the ac-cident was not an anomalous, random, or "normal" event, but rather one rooted in NASA's history and the human space flight program's culture. While the physical cause of the loss of *Columbia* and its crew was a breach in the thermal protection system due to a piece of insulating foam that sep-arated during the launch, the organizational causes stemmed from years of resource constraints, scheduling pressures to launch, and what the board characterized as "cultural traits and organizational practices detrimental to safety."[9] After criticizing NASA managers for their overreliance on past success and establishment of organizational barriers that stifled professional differences of opinion and prevented the communication of critical safety information, the board concluded that "NASA's organizational culture had as much to do with this accident as the foam did."[10]

Overreliance on past success, like the normalization of deviance, is found in studies of risk management in organizations. In some organizations, a culture develops that is not necessarily sinister, but that downplays risks. When organizations engaged in high-risk technologies have a number of close calls but never a disaster, they may become susceptible to what William Freudenburg and Robert Gramling refer to as the "atrophy of vigilance."[11] Over time, organizations lucky enough to remain accident-free become overconfident in their ability to avoid problems, and that overconfidence leads eventually to a failure to pay attention to the precursors of a catastrophe.

However, atrophy of vigilance and organizational normalization of deviance do not fully address why large, highly profitable companies engage in wanton misconduct. We might argue that struggling companies on the brink of failure have a much stronger incentive to cut corners, owing to competition and their desire to survive, and that well-heeled corporations with healthy balance sheets, being better able to include the costs of compliance with environmental and safety regulations in their costs of doing business, are less likely to misbehave. But that is not always the case.

Consider Tyco, a company specializing in fire and safety equipment; Enron, an energy trading company; and WorldCom, a telecommunications company. These companies were darlings of Wall Street for a time—each experienced meteoric rises in stock price and handsomely paid its executives. Nevertheless, the executives of these companies were convicted of federal crimes between 2001 and 2006 and are currently serving time in prison. (The exception is Kenneth Lay, chief executive at Enron, who died before he could serve his forty-five-year term.) The companies fared little better. Enron and WorldCom filed for bankruptcy, Tyco split into separate companies, and employees and shareholders paid dearly for their trust in these organizations.

So some highly profitable companies decide to go to the "dark side"—beyond the kind of organizational cultural issues found in the NASA examples—even though they can afford to comply with regulations, and even as they know the right thing to do. Company executives such as those found at Enron or Tyco act with callous disregard for safety protocols and the environment. Why? One answer to this puzzle involves organizational power and executives' hubris.

James Reason notes that the most grievous errors in high-technology enterprises come not from the frontline operators but from the "blunt end" of the system—that is to say, the high-level decision-makers.[12] He suggests that the further these individuals are from frontline activities and potential accidents, the greater the potential danger they pose to the system. When top-level managers suggest to workers that safety and environmental regulations are overblown or costly, they imply that they expect a certain kind of conduct from frontline personnel, and that expectation becomes part of the fabric of the organizational culture. So, too, do internal performance pressures to deliver products and services as quickly and efficiently as possible. In turn, these pressures may affect individual actions and foster the development of an internal culture that implicitly supports achieving organizational goals in illegitimate ways.

Such expectations are unlikely to support worker positions that run contrary to the organizational way of thinking. Ultimately, top administrators may place performance pressures on staff either indirectly—for example, by establishing out-of-reach goals or by not providing sufficient resources to attain goals—or directly if they foster a climate that supports misconduct as a way of reaching the organization's goals.[13] Either kind of pressure (or both) sets up an organization for a shift to the "dark side." In voicing his concern about BP's drive to finish the Macondo well in the days that led up to the disaster in the Gulf of Mexico, William Reilly, cochair of the National Commission on the BP *Deepwater Horizon* Oil Spill and Offshore Drilling, noted that "a safety culture must be led from the top, and permeate a company."[14]

In examining why good companies do bad things, Peter Schwarz and Blair Gibb also identify problems at the top. They find that executives and managers often will not tolerate dissent from their workers. Rather than encourage feedback or differing opinions, executives tend to isolate themselves, talking to the same people and using the same sources of information. They let their commitment to getting a project completed overwhelm any other consideration as they focus exclusively on financial indicators of performance.[15]

Power is the lifeblood of organizations, and individuals—managers and top executives—are not immune to the heady allure of control, whether in public or private organizations.[16] A company's success and prominence may lead to hubris on the part of its executives and managers. If success is ongoing, managers may see themselves as infallible and become more willing to take risks and less willing to see the pitfalls of potential actions.[17] They may also seek to maintain control, avoid operational transparency, and ignore recommendations that do not lend themselves to the pursuit of profit at all costs. Hubris coupled with power makes top-level executives and managers more inclined to believe that they can outsmart regulators and skirt regulatory requirements while pleasing stockholders.

It's worth emphasizing here that public and nonprofit organizations, such as NASA in the *Challenger* and *Columbia* accidents, are not immune to these issues. More recently, the case of the lead in the drinking water in Flint, Michigan, shows some elements of poor decision-making and disregard for safety and environmental concerns on the part of public managers. Flint's drinking water became contaminated with lead in 2014, while the city was under the control of a state-appointed emergency manager who had authorized switching from water provided by Detroit's water system to raw water taken from the Flint River. Since the raw water was not properly treated, it leached lead from water pipes. Subsequently, the Centers for Disease Control and Prevention (CDC) found that children in Flint had almost a 50 percent higher chance of elevated blood lead levels, which threatened their health and mental development. As of December 2016, forty-eight criminal charges had been filed against thirteen state and local officials who played a role in allowing the lead to leach into the city's water supply. The two emergency managers for Flint, Gerald Ambrose and Darnell Early, who reported directly to the governor of Michigan, were charged with multiple felony counts.[18]

In sum, organizational culture is influenced by the people who lead and manage the organization. They can create a culture that supports safety and environmental stewardship, or one that reorients the attention of staff away from potential hazards and toward producing more, finishing the task at hand, and perversely valuing an ignorance of what might go wrong. The drip-drip-drip of waning vigilance in the latter type of organization continues until disaster strikes.

Four Stories and Four Common Characteristics

Each of the four stories in this book depicts an overt industrial disaster in which a company's actions resulted in a loss of life and significant environmental degradation. Each disaster story describes a similar suite of bad decisions, summarized in the four common characteristics of these events: (1) a history of complacency and disregard for safety and environmental

> **BOX 1.1:** Four Common Characteristics of Industrial Disasters
>
> 1. An organizational history of complacency and disregard for safety and environmental regulations
>
> 2. A narrow-minded focus by leaders and managers on maximizing profits at all costs
>
> 3. Inadequate planning and preparation for addressing safety or environmental conditions when they arise
>
> 4. A political environment that encourages lax enforcement of safety and environmental regulations

regulations; (2) a focus on company profitability at all costs; (3) inadequate planning and preparation for accidents; and (4) a political environment in which regulators are pressured not to enforce regulations (see Box 1.1). The next four chapters will describe these characteristics in greater detail, but here I'll provide a short overview.

A History of Complacency and Disregard for Safety and Environmental Regulations

It is perhaps the most important characteristic shared by these four industrial disasters that they all had a history of complacency and disregard for safety and environmental regulations. Chapter 3 traces the story of the explosion of BP's *Deepwater Horizon* oil rig. The special presidential commission investigating that disaster charged that BP had engaged in a "culture of complacency" regarding safety protocols.[19] The company settled with the US Justice Department in 2015 for $20.8 billion—the largest civil penalty in the history of environmental law.[20]

Perhaps unsurprisingly, a less than stellar corporate history plagues the company. Prior to the 2010 Gulf of Mexico oil spill, BP had admitted to breaking US environmental and safety laws and committing fraud and had paid nearly $373 million in fines to the US government.[21] This descent into poor judgment included instances of mismanagement in 2002 at its facility in Prudhoe Bay, Alaska; the massive explosion in 2005 at its refinery in Texas City, Texas, which killed 15 workers and injured 180; and a spill of more than 200,000 gallons of crude oil on Alaska's North Slope in 2006.

Chapter 2 tells the story of how thousands lost their lives in what is known as "the Night of the Gas": the explosion of a Union Carbide pesticides facility in Bhopal, India, on December 3, 1984. Like BP, Union Carbide did not follow proper protocols in caring for its methyl isocyanate gas, nor did it take precautions to protect the community, even though warning signs in the system were present at the time of the explosion and indeed had been present for months. Like their BP counterparts, Union Carbide officials were lackluster at best in monitoring the company's operations.

Massey Energy, headed by a strong-willed CEO, Don Blankenship, failed to properly vent and dust the Upper Big Branch mine (see Chapter 4). It exploded on April 5, 2010, killing twenty-nine miners. In the aftermath of the disaster, federal agents learned that the company had covered up safety violations rather than deal with them. The Zonolite vermiculite mine owned by W. R. Grace in Libby, Montana, offers yet another story of a company that exhibited a lack of concern for safety and health regulations (see Chapter 5). Not only were workers at this mine exposed to tremolite asbestos fibers, but so, too, were their families and even people who had no connection to the vermiculite operation. Executives at W. R. Grace faced charges of wire fraud, obstruction of justice, conspiracy, and violations of the Clean Air Act (CAA) of 1970. Although the company and three of its executives were acquitted of all charges in 2009, internal company memos revealed that mine officials knew that asbestos fibers were in the vermiculite coming from the mine.

A Focus on Company Profitability at All Costs

Another characteristic shared by these four industrial disasters is the focus of the companies on profitability over anything else. It appears that BP deliberately cut corners in an effort to meet its completion deadline for the Macondo well and keep costs down as much as possible. At the time of the spill, BP was nearly $43 million over budget and forty-three days behind schedule, and so the company had chosen to use cheaper construction and materials and to eliminate time- and money-consuming safety precautions and tests.[22] In Bhopal, the Union Carbide plant, hemorrhaging money because of stiff competition from a safer insecticide method, had put in place draconian cost-saving measures, which included cutting in half the workforce who managed the methyl isocyanate gas unit and shutting off the refrigeration needed to keep the gas cool and prevent it from vaporizing or reacting.

It is perhaps fitting that their attention to profitability did not serve these companies in the aftermath of the tragedy. Two of the companies, Massey and Union Carbide, were bought out after these catastrophic events. W. R. Grace sought bankruptcy protection. BP lost nearly $90 billion in market capitalization in the months after the explosion and was forced to sell some facilities to cover cleanup costs. But the far higher costs were the human and environmental toll of the disasters.

Inadequate Planning and Preparation for an Accident

Inadequate contingency planning is closely linked to the pursuit of profits and revenue generation. Companies in high-risk enterprises need to ensure that their investments in safety and accident prevention are comparable to their investments in other areas. Regulators, in turn, must ensure that these response plans exist, that they comply with regulatory requirements, and that they are sufficient in scope to cover a wide array of accidents. Unfortunately, in the four industrial disasters discussed in this book, planning and preparation were sadly lacking.

BP, for example, was subject to the requirement under the Oil Pollution Act of 1990—which Congress passed in the aftermath of the 1989 *Exxon Valdez* oil spill in Alaska—that oil companies prepare oil spill response plans. Regulations of the US Department of the Interior's Minerals Management Service (MMS) require that the plans include details about emergency response actions, procedures to be followed in case of a spill, and a calculation of (and planning for) the worst-case scenario. BP's worst-case scenarios for an oil spill at the Macondo well ranged from 28,000 to 250,000 barrels of crude oil—far from the more than 3.19 million barrels that eventually poured into the Gulf of Mexico. Equally troubling, the response plan was so perfunctory in its approach, drawing from previous plans and even government websites, that it incorrectly listed several marine species that did not exist in the Gulf, such as sea lions and walruses.[23] According to a government investigation following the disaster, the MMS approved the plan without additional analysis.[24]

Massey Energy also neglected to adequately prepare for the disaster at the Upper Big Branch mine. Mine Safety and Health Administration (MSHA) regulations require mine operators to examine mines on a regular basis, a task assigned to the company's mine examiners. However, testimony during the MSHA investigation revealed that Massey inadequately trained its examiners, foremen, and miners in how to recognize and address a hazard. So intent was the company on hiding its dereliction of responsibility to protect workers at the mine that MSHA inspectors were shown a different set of the required examination books. The examination books shown to federal inspectors lacked accident data as well as identified hazards. At the time of the explosion, the hazards described in the hidden second set of books remained unaddressed.[25]

The other two stories reveal a similar lack of preparation for accidents and emergency response, as illustrated in the chapters that follow.

A Political Environment That Fosters Cozy Relationships with Regulatory Personnel and Government Officials

The companies and their actions were not the only contributors to these industrial disasters. Lack of oversight from regulatory personnel and government officials played a part as well. In the BP case, the federal agency responsible for ensuring that an offshore drilling rig was operating safely, the Minerals Management Service (MMS), fell short. Studies revealed that the MMS conducted fewer inspections of rigs and allowed hundreds of drilling plans to move forward without sufficient oversight or even the required environmental permits. As Kieran Suckling, director of the environmental organization Center for Biological Diversity, observed in the weeks following the spill: "MMS has given up any pretense of regulating the offshore oil industry. . . . The agency seems to think its mission is to help the oil industry evade environmental laws."[26] The Obama administration would reorganize the agency that year.

In Union Carbide's case, the Indian government had encouraged the chemical manufacturer to locate in Bhopal, hoping to stimulate local economies and provide materials for its agricultural activities. Local officials knew about the toxic nature of the chemicals at the plant and the growth of the shantytowns around the plant, and they had access to technical documents and reports warning of impending dangers, and yet they did nothing. Indeed, a government official sheltered Warren Anderson, the CEO of Union Carbide, when he came to visit the plant in the days after the tragedy, thus helping Anderson avoid arrest by the Indian government.

For both the Upper Big Branch and the Zonolite mines, federal and state inspectors were not as determined to protect the miners as they should have been, as will be described in subsequent chapters. The MSHA admitted as much in noting that in the six years after the West Virginia mine explosion, the agency became "ever vigilant in enforcing the regulations designed to keep miners safe," having conducted 1,113 impact inspections and issued over 15,000 citations at mines that "merit increased agency attention and enforcement" between 2010 and 2016.[27]

High-Reliability Organizations

In all four industrial disasters, environmental and public health tragedies might have been avoided altogether, or at least minimized. High-reliability organizations (HROs) aim to do just that.

High-reliability organizations consistently achieve extraordinary levels of reliable performance. They are exceptionally consistent in avoiding catastrophic errors, even though they have equally complex and tightly coupled systems that have the potential to create the system accidents described by Perrow. These organizations achieve reliability and avoid sliding to the "dark side," however, and one way they do so is by focusing on technical expertise, and establishing a culture that values staff input.

Studies of such organizations by Todd La Porte, Karl Weick, Kathleen Sutcliffe, and their colleagues have shown that HROs beat the odds for having "normal" accidents by embracing common characteristics and values.[28] Weick and Sutcliffe found that five principles guide HROs: (1) a preoccupation with failure; (2) a reluctance to simplify interpretations; (3) a sensitivity to operations; (4) a commitment to resilience; and (5) a deference to expertise (see Box 1.2).[29] These principles stand in sharp contrast to the characteristics of the four companies responsible for the disasters related in this book (Box 1.1).

First, in their preoccupation with failure, HROs incorporate safety into their core mission and their culture. They do not just make a token commitment to protecting the environment and public and employee safety—they put this commitment front and center in their processes and operations.

Second, staff in HROs are reluctant to simplify a problem or its solutions. In contrast to the atrophy of vigilance displayed by BP, Union Carbide, Massey Energy, and W. R. Grace, they remain attentive to conditions that could cause accidents.

Third, HROs are sensitive to operations, such that employees pay close attention to what is and isn't working and probe to more deeply understand problems as they arise. As Weick and Sutcliffe put it: "Reliable performance tends to increase when close calls are interpreted as danger in the guise of safety and to decrease when close calls are deemed as safety in the guise of danger. . . . Operations are in jeopardy when their soundness is overestimated. When people see a near miss as success, this reinforces their beliefs that current operations are sufficient to forestall unintended consequences."[30]

Fourth, because HROs are aware of the risks associated with their work and the need to think resiliently, staff are rewarded, not penalized, when they discover and report errors or safety concerns. HROs develop an intrinsic ability to maintain a stable state, even when errors occur. They bounce back from small errors and learn from them.

BOX 1.2: Five Principles of High-Reliability Organizations

1. **Preoccupation with failure:** Focusing on what could go wrong and remaining vigilant when systems or processes go awry

2. **Reluctance to simplify interpretations:** Wanting to get at the root cause of a problem and engaging in careful analysis to find solutions

3. **Sensitivity to operations:** Being aware of organizational changes as well as changes in circumstances or situations; developing mindfulness in the organization

4. **Commitment to resilience:** Paying attention to the skills and talents of people in the organization and dedicating resources to careful planning and training

5. **Deference to expertise:** Believing that staff on the front lines are experts and listening to their suggestions and concerns regarding safety protocols

Source: Karl E. Weick and Kathleen M. Sutcliffe, *Managing the Unexpected: Sustained Performance in a Complex World* (San Francisco: Jossey-Bass, 2015).

Fifth, HROs' commitment to mindfulness and organizational attention involves a strong sense of teamwork, flexible decision-making, and decentralized power arrangements. HROs strive to break down the power silos of senior executives by ensuring that personnel have ready access to senior management and that safety and environmental professionals are placed high in the organizational structure. At the same time, HRO leaders defer to expertise wherever it is found in the organization, regardless of hierarchy or seniority. To put it simply, leaders listen to people who have the deepest knowledge of the task at hand.

In the highly technical and technological environment of our complex world, many organizations are prone to catastrophic failures. Avoidable accidents tend to occur in organizational cultures with executive hubris, normalization of deviance, atrophy of vigilance, micromanagement, a fixation on profit or project, and a disregard for the warning signs of system failures. By contrast, companies with cultures that seek to minimize risk and build resiliency, that trust subordinates to bring information forward, and that encourage constant learning are more likely to avoid catastrophic failures.

Now that we have considered the qualities within an organization that might either lead to or prevent disasters, let us turn to the role of external factors in shaping organizations and the need for regulations.

THE ROLE OF GOVERNMENT IN SHAPING ENVIRONMENTAL STORIES

External Pressures and the Need for Regulation

Organizations are not immune to external pressures, including the high expectations placed on them by stakeholders, including stockholders, governing boards, regulatory agencies, clients, customers, and citizens. External pressures take several forms. In a corporate setting, the company may face increasing pressure to perform well and bring high returns to stockholders. Yuri Mishina and his colleagues have found that prominent, highly profitable companies may be more likely to turn to illegal activity, because of the market's ever-increasing expectations for their future returns.[31] Corporate CEOs and top executives, they find, perceive increasing pressure to outperform competing firms in order to command premium prices in the stock market. The prospect of poor future relative performance may compel high-performing firms to engage in illegal activities in order to maintain their prominence in American and international business venues. Mishina and his colleagues warn: "Regulators should endeavor to monitor the activities of both high- and low-performing firms to detect illegal corporate behavior, and they should consider a firm's prominence and performance relative to industry peers in assessing which firms should receive closer attention."[32]

Behavioral economics suggests that because companies seek to maximize profit, they will reduce costs in areas where they can. This self-interested behavior on the part of businesses establishes the need for government regulation in social policy areas such as the environment, workplace safety, and consumer protection. Regulations establish the ground rules for operating in an economic system where everyone has access to common resources, such as air. In the absence of regulation, businesses, with an eye toward their bottom line, will ignore or disregard the environmental consequences of their activities, preferring to exclude the costs of pollution from the cost of production. As Garrett Hardin famously quipped in his essay "The Tragedy of the Commons," "freedom in a commons brings ruin to all."[33] Laws such as the Clean Air Act and the Clean Water Act were designed in part to force companies to account for the costs of minimizing pollution in their cost of doing business. Laws establish permitting requirements and emission limitations and are followed by hundreds of regulatory standards set by the states and the US Environmental Protection Agency (EPA). This is a way of accounting for the otherwise predictable behavior of businesses and individuals as rational actors.[34] Absent regulation, businesses have little or no incentive to control pollution or to use natural resources efficiently.

Even with regulation, companies may engage in a compliance calculus. If companies perceive that the probability of being caught violating regulatory requirements is small, or that enforcement by regulators is weak, they may simply choose to ignore those requirements and slip ever closer to the "dark side." "Command-and-control" regulatory systems such as those found in most major US environmental laws work best as long as the regulations have a reasonable expectation of being implemented and businesses sense that either on-site inspections or self-reporting requirements could lead to heavy sanctions. The extent to which they are willing to engage in illegal behavior or criminal misconduct may be explained, in part, by this kind of compliance calculation.

A second form of external pressure comes from expectations put on an organization to perform in other ways. Policy scholars who study disasters suggest that failures in complex systems must be understood within not only an organizational context but also the political, economic, societal, and cultural context in which they operate.[35] For example, NASA was under political pressure to deliver a launch or face funding cuts from Congress. Societal demands for goods and services may prompt companies to grow too fast or to embrace unproven technologies. Part of the story of Union Carbide in India can be traced back to political and societal demands. Famine in India in the 1940s had killed millions of people, and in an effort to avoid future food shortages, India embraced the Green Revolution (roughly 1967–1977). In its effort to move from being a food-deficient country to a self-sufficient agricultural nation, India responded to the increasing demand for pesticides by embracing US companies like Union Carbide that could provide agrochemical-based pest and weed controls.

The politics favoring an industry may also influence the behavior of a particular organization. The political environment may be very forgiving of a company that meets the public's energy or food needs, even to the point of calling for a relaxing of regulations. These political forces may, in turn, undercut the ability of a regulatory agency to perform by reducing budgets, which may compel reductions in inspection and enforcement personnel, or through political appointments that influence the scope of regulatory pressure on a particular industry. US political winds at the time of the BP oil spill supported domestic oil production and offshore oil drilling. Politicians heralded BP's deepwater drilling, seeing it as a boon for energy security and domestic energy production. The MMS at the time of the spill was underfunded, faced political pressure to open the Gulf for drilling, and was armed with insufficient rules and regulations to properly oversee the *Deepwater Horizon*. After the BP spill, new offshore drilling leases in the Gulf were temporarily suspended, but existing drilling continued until new regulations were put in place.

In short, we can look inside an organization to determine the extent to which it has instilled HRO practices and culture, but a full story also accounts for external social, economic, and political pressures on the industry as well as on the regulatory agency. This fuller story compels us to examine the socialization of risk—the spreading of risk across the public at large. As Perrow puts it:

> A focus upon a culture of reliability is a luxury in the world of risky systems, one that I hope we can afford, and one the social science disciplines such as psychology, sociology and political science can address and profit from. *But an economic system that runs such risks for the sake of private profits, or a government system that runs them for the sake of national prestige, patronage, or personal power, is the more important focus and culprit. . . . The issue is not risk, but power.*[36]

Responding to Disasters: Agenda-Setting for Laws, Policies, and Regulations

Each of the disaster stories in the book explores how environmental laws, regulations, and policies were either created or changed by how situations unfolded. How and when government responds to public issues has long captured the attention of policy scholars. "Agenda-setting" is an early step in the policymaking process and refers to moving the issue or problem from the public sphere into a formal process through which it can be addressed by policymakers. For example, an issue is put on the congressional agenda when a bill is introduced. Agenda-setting can be further conceptualized as comprising three elements: the media agenda, the public agenda, and the policy agenda.[37] In these stories, the main focus is on the policy agenda. However, the media agenda (what is covered in the news, how it is covered, and for how long) and the public agenda (what we are paying attention to) often act in symbiotic ways and play a role in shaping the policy agenda—or what policymakers attend to.

Thomas Birkland posits that political reactions to disasters and the subsequent influence on the institutional agenda depend on the nature of the event. If the event is big, it captures a lot of media attention. Disasters also mobilize groups that advocate for new laws, regulations, or other governmental responses and that use the media coverage to gain support for their positions. A powerful tool for such advocacy groups may be the use of symbols from the disaster that exact emotional responses from the public.[38] Many of us remember the oil-covered sea otters and birds that became emblematic of the problems of oil spills after the *Exxon Valdez* ran aground off the coast of Alaska in 1989. These powerful images helped spur the quick passage of the Oil Pollution Act in 1990.

The media alert people to an issue and provide the context for setting the public agenda by choosing how to frame that issue.[39] In helping people become aware of a problem and setting the context for how we understand it, the media help place the issue on the public agenda.[40] However, as media attention to an issue wanes, so too often does public attention. Thus, issues on the public agenda tend to be displaced quickly by other issues. As Anthony Downs notes, "American public opinion rarely remains sharply focused on any one domestic issue for very long—even if it involves a continuing problem of crucial importance to society."[41] Thus, issues rapidly recede from public view, the public mind, the media focus, and the attention of policymakers in what Downs describes as the "issue-attention cycle."

The bottom line is that issues have a relatively short time frame in which to be transferred onto the policy agenda, where government responds by creating new laws, changing existing laws, appropriating funding to address the problem, or taking action in government agencies or through executive actions. This creates an environment of conflict and competition for those who want policymakers to pay attention to their issues.[42] It also creates a special role for events that capture media and public attention. A night of watching the news will suggest what researchers have documented: media outlets value "newness" and sensational stories and are thus biased toward issues that emerge onto the public stage in dramatic ways. These events crystallize attention and can powerfully shape the likelihood of a problem being attended to by policymakers.

Issues that catapult onto the agenda because they are event-driven— that is to say, by a tragedy, an unexpected event, or an emergency—tend to gain the attention of media and policymakers. In turn, these triggering, or focusing, events can push an issue onto all three agendas: the media, public, and policy agendas.[43] Reaching the policy agenda, however, does not guarantee that new laws or policies will result. It is one thing for policymakers to talk about an issue; it is quite another for them to actually do something about it. After all, the media deals in immediacy, while policymakers are more likely to deal in contemplation, deliberation, and debate. Every decision that results in a new law or policy will, in turn, arouse some level of

tension during implementation. Thus, members of Congress or other policymaking venues, such as state legislatures, seldom act with great speed,
if at all.

In studying the ability of an issue to reach the policy agenda and stay on
it long enough to result in formal responses, John Kingdon theorized that
agenda-setting is best modeled as the confluence of problem identification
and recognition (often seen through the lens of a triggering event), the avail-
ability of solutions to solve the problem, and the politics surrounding policymaking, which he notes include the "national mood, vagaries of public
opinion, election results . . . and interest group pressure campaigns."[44] Thus,
agenda-setting requires more than just the propelling force of a triggering
event, though that is important. Ultimately, Kingdon's "multiple streams"
theory also requires the existence of viable alternatives and the subsequent
narrowing of these alternatives to solutions that are publicly and politically
acceptable. With no acceptable alternatives or the political will to adopt
them, the problem will quickly be replaced by other compelling issues that
may have viable solutions.

The Bhopal and BP cases, in particular, brought about changes in environmental policies and laws. Their appearance on the policy agenda and the
way they captured media and public attention reflect many of the elements
of the agenda-setting process described here.

Responding to Disasters: Policy Implementation and Policy Change

Each story in Chapters 2 through 5 involves not just agenda-setting but also
policy implementation and policy change. Three of them—the asbestos disaster in Libby, Montana, the BP oil spill, and the explosion at the Upper Big
Branch mine—offer valuable insights into how environmental laws and regulations are implemented. Policy implementation happens once a policy is
formalized (such as when a law is passed). Federal or state agencies are given
the challenging task of exacting behavioral change from the target group
identified in the policy. So, for example, the Minerals Management Service
was charged with implementing US policies and regulations on offshore oil
drilling. The EPA is tasked with implementing, or putting into practice,
many of our environmental laws.

Measuring implementation is a matter of asking: What happened after
the policy was formulated? And why did it happen this way? Implementing
a public program involves multiple actors involved in a complex process that
changes over time. Successful implementation depends on a number of factors, including the nature of the problem that caused the policy to be created

in the first place, political support for the program, the role orientation of street-level implementers, court interpretations of statutory language, and the congruence of the policy goals with those of the implementing agency.

Existing policies often change as a result of a disaster or a major event. Policy scholars have not determined the exact relationship between disasters and changes in policy, but they suggest that such changes depend on how the event is interpreted by policymakers. Interpretation, in turn, is shaped by how the disaster is framed—as explanations emerge to explain the disaster, the one with staying power in the public domain is likely to shape the nature of the change. Additionally, as policymakers learn about the causes of the disaster, they may alter their beliefs about the nature of the operation. For example, the Mine Safety and Health Administration became more vigilant and changed its approach to inspecting mines after the Upper Big Branch disaster.

Frank Baumgartner and Bryan Jones developed a theory of "punctuated equilibrium" to explain why some policies languish for years, with little or no change, but then, on rare occasions, become susceptible to major change.[45] They felt that while stasis, or incrementalism, generally explains most policy changes, there can be times of rapid, dramatic shifts in policies, such as during a crisis. Most policies develop into what Baumgartner, Jones, and Peter Mortensen characterize as monopolies that maintain the status quo.[46] Large punctuations, or policy shifts, occur in part because entrenched political allegiances are shaken by different ways of viewing the policy and its implementation. Policymakers are drawn to understand the defects in existing policy approaches to the public problem as new policy "images" emerge that challenge existing ways of understanding that problem. One example of punctuated equilibrium is found in America's approach to asbestos, which is described in Chapter 5. Here we see little or no change in policies regulating asbestos until waves of people who had worked in industrial settings, mostly during World War II, became ill with asbestosis, mesothelioma, and lung cancer, prompting a tsunami of litigation, new laws, and new asbestos regulations.[47]

In sum, accidents and tragedies are part of the human experience. Unanticipated consequences are woven into our societal fabric, and we should expect that bad things will happen, especially in complex systems and high-risk enterprises. At the same time, it is appropriate to consider the values, intent, and motivations that individuals bring to their organizations. Exploring the culture permeating an organization may help us understand the actions taken by individuals with decision-making authority in that organization. Failures in complex systems should also be understood not only within an organizational context but also within the political, economic, societal, and cultural situation in which they operate.

Government plays a critical role in helping to protect the environment and its citizens. External pressures, as described earlier, exist for all organizations, and public institutions, while striving to protect the common good, face political pressures as well. As issues move onto the formal agenda, often spurred by an event such as an industrial disaster, governments must decide how to respond. Many environmental laws, including the ones relevant to our four stories, require that regulatory agencies serve as watchdogs over companies, promulgate environmental standards, and then implement those standards. Implementing policies is a key task of federal and state government agencies. In the best-case scenario, policymakers learn from their success, or lack thereof, in implementing environmental policies and make changes that help address emerging issues.

VILLAINS AND HEROES

The previous section deals mostly with organizational behavior and external forces that create conditions for these organizations. This section focuses on the role of *individuals*. Heroes and villains are characters in most good stories, and this is true for many environmental stories as well. At the same time, it's important to view heroes and villains on a continuum: at one end are the purest heroes, and the most egregious villains reside at the other end. Most storybook heroes are flawed, possessing some bad qualities. By the same token, few villains are completely contemptible.

Whether to identify someone as a villain rests in the eye of the beholder, but certain characteristics may help sharpen the view. Top-level executives and managers of organizations capture our attention and are most likely to be seen on the villainous side of the continuum when things go badly. Certainly, they are spotlighted in the media and in the courts as responsible for the actions of their organization. As the face of their organization, they are the ones who are blamed for accidents or misconduct that leads to disaster, and rightly so in many situations, as managers and executives set the tone for the organization, as discussed in the previous sections. They establish the incentives that, in turn, motivate employee response. They make decisions to be sustainable, or not, and they decide whether to prioritize safety as a key element of the organizational culture, or not. Box 1.3 presents a list of characteristics that help define environmental villains and heroes.

The first characteristic of an environmental villain at the far end of the continuum is a prolonged and consistent disregard for, or discounting of,

BOX 1.3: Characteristics of Environmental Villains and Heroes

Environmental Villains	Environmental Heroes
• Consistently make choices that do not protect the environment, human health, or safety	• Consistently make choices that are ethical and seek to advance the greater environmental and social good
• Act with knowledge and with intentional disregard for potential environmental or human health risks	• Are indefatigable in standing on principle in tough times and define a new standard of excellent conduct and integrity
• Are powerful, influential, and in possession of resources sufficient to act	• Pursue an environmentally responsible task or objective regardless of resources available; never give up
• Obfuscate the truth and have below-board agendas that ultimately put the environment or people in peril	• Demonstrate courage in pursuing environmental goals in the face of powerful antagonists
• Are prideful and seek to distance themselves from responsibility for an accident or disaster	• Exercise wisdom in pursuit of environmental goals; often credit others for accomplishments
• Exhibit little regard for the well-being or opinions of others	• Believe that people acting together can bring positive change

potential harm to environmental and human health. Environmental villains also act with knowledge. They understand that risks to the environment or to human health and safety exist, sometimes for years, yet choose, with knowledge, to do nothing. If heroes are courageous, principled, wise, and noble individuals, villains are "deliberate scoundrels or criminals," as the Merriam-Webster's dictionary defines them. In short, they know better, they understand that their lack of action places the environment and people at risk, but they still do nothing to alter their course. Thus, villains cannot really be surprised when tragedy hits because they knew all along that the potential for harm was being magnified by their actions.

Another villainous personality trait is the ability to accumulate power and influence and use it in perverse ways. More to the point, the villains who best capture our attention appear larger than life. They control people and systems and answer to no one. Villains also obfuscate the truth. In the environmental arena, examples include hiding safety issues from federal or state officials to avoid charges of violating environmental or occupational safety laws, or pressuring environmental officials to look the other way when problems are evident. Sometimes this behavior is subtle and hard to detect; other times it is blatant, in full view of the public. Take, for example, Don Blankenship, the former CEO of Massey Energy, who was widely regarded as a micromanager and who instructed coal mine managers to "run coal" instead of taking the time to fix safety issues (Chapter 4).

Finally, after disaster strikes, environmental villains seek to shift blame away from themselves. They minimize the extent of their control over others in an organization. Top-level executives frequently blame employees or midlevel officials for the environmental harm resulting from their own devious business practices.

Now for the other side of the continuum—the heroes. Everyone loves heroes. We look for them to save us from the "bad guys" and to virtuously lead us in the right direction, even when we may not want to go down that path. Some of our comic book and movie heroes possess superhuman abilities, like Peter Parker's "spidey" sense, while others rely more on their intellect, like Batman with his high-tech gadgets. They all act with a moral compass that keeps them from giving in to appeals by the forces of evil or giving up when the going gets tough. Superheroes shun the spotlight and seldom falter.

In looking at the philosophical foundations of heroism, Andrew Bernstein identifies four components of heroism: moral greatness, ability or prowess; action in the face of opposition; and triumph in at least a spiritual, if not a physical, form.[48] This spiritual triumph implies an unbreakable allegiance to values that advance humankind. Bernstein notes that heroes pursue goals indefatigably in the face of powerful antagonists. It is persistence and fierce defense of "the good" that defines a hero—even if the hero ultimately fails against the villain. A hero is fundamentally defined by this integrity and nobility of character. As put by Judy Logan and Gail Evenari in undertaking the Heroism Project, "In principle at least, heroes represent the finest qualities of our collective character."[49]

Ability need not be physical. Heroes certainly emerge on battlefields as they vanquish the enemy, but they also are found in ordinary and everyday experiences, especially those that require careful thought in advancing the greater good. Indeed, heroism requires application of knowledge, rational thought, and the exercise of good judgment. Heroes are thinkers—underlying their actions is their wisdom.

But the key distinction between a hero and a role model or an otherwise moral person is courage. Courage implies conflict and adversaries. In an easy situation that involves no fear and no risk, there can be no heroism.

Remaining dedicated to what is right even in the face of threats, however, requires courage. Those threats may come in the form of physical dangers, risk of criticism or vitriol by certain members of a community, loss of livelihood, or other kinds of physical, mental, or financial sacrifice. As Bernstein observes, "Courage is integrity in a context: it is unyielding commitment to one's values in the teeth of a force or foe that threatens them. The brave man is not necessarily one who is unafraid but one who performs whatever protective actions his values require, no matter the intensity of his fear. This bravery is the especial moral hallmark of the hero."[50]

Resourcefulness is another defining characteristic of heroes. Heroes take whatever they have to work with and try to make it into something extraordinary. In their best-selling work on leadership, James Kouzes and Barry Posner posit that successful leaders consistently challenge the process and not only look for opportunities to do things better but are also willing to take risks to improve organizations.[51] In studying federal employees, Norma Riccucci discovered many positive change agents, even in agencies where budgets were cut or politicians attempted to shift the mission away from the public good to serve more specialized interests.[52] In her words, these individuals were "unsung heroes" who embraced an ethos of ingenuity and a sense of purpose. Heroes, unsung or well known, leverage what they have to make something happen.

Environmental heroes employ their heroic qualities in pursuit of what they understand to be a greater environmental good. They share a deep commitment to the environment, whether that is a watershed, river, community, state, region, or country, or the entire planet. Dedicated to their environmental values, they often act in the face of conflict, confronting foes with much greater resources. They act with integrity, with courage, and with wisdom. A key characteristic of environmental heroes is their belief that people acting together can be the catalyst for positive change. Heroes may lead, but they do so with the confidence that others will follow. They embrace an environmental, or ecological, view of citizenship that acknowledges the willingness of others to take on the responsibility for protecting the earth. Heroes motivate "the rest of us" to be better environmental stewards, as illustrated in Chapter 6.

CIVIC ENVIRONMENTALISM: THE STORY OF THE REST OF US

If you've read this far, you may be wondering what all of this has to do with you. In a word: everything. We make choices every day that have environmental consequences. Do we use renewable sources of energy? Do we consider fuel efficiency when buying a vehicle, or do we take public transportation? Do we recycle, turn off the lights, minimize our use of plastics, such as plastic bags and water bottles, and conserve water? What each of us chooses to consume and how we choose to behave matter.

Perhaps more important is how we engage in the political process. On April 22, 2017, hundreds of thousands of people participated in the March for Science, held on Earth Day. The march represented a political response to what the citizens participating viewed as egregious actions by the Trump administration to dismantle science agencies (such as the EPA) and remove climate data from government websites. A week later, the People's March for Climate Change drew thousands more people to Washington, DC, and satellite marches around the country and the world. The differences in approach to science and climate change between the Obama and Trump administrations suggest that political participation will shape the future directions of public policy. If environmental protection is important to us, it should be reflected in our voting decisions, as well as in other ways that we participate in politics, such as contacting our elected representatives in local, state, and national government. In doing so, we become part of a long history of environmental activism.

The history of US environmental and conservation movements is filled with stories of individuals who fought to protect wilderness areas, establish parks, protect rivers and watersheds, and warn us about the dangers associated with chemicals such as pesticides. Books have been written about many of these individuals who so powerfully shaped our environmental heritage or helped us to understand the need for tough new national environmental laws—John Muir, Teddy Roosevelt, Aldo Leopold, Rachel Carson, Lois Gibbs, Ralph Nader, just to name a few. There are also celebrities who speak passionately about the environment and have dedicated their time and resources to environmental organizations and causes—the actors Leonardo DiCaprio, Mark Ruffalo, Daryl Hannah, Ed Begley Jr., and Robert Redford, for example.

Elected officials have also done yeoman work to call attention to environmental issues, thus influencing the course of the American environmental position; Edmund Muskie, Al Gore, Barbara Boxer, Henry Waxman, John Chafee, John Kerry, and Robert Kennedy Jr. are just a few national-level politicians who come to mind. Hundreds of staff in local, state, and national agencies have dedicated their careers to establishing and advancing environmental programs. Leaders of environmental organizations, such as Edgar Wayburn and David Brower of the Sierra Club, Bill McKibben who founded 350.org, and Vandana Shiva who founded the Navdanya Research Foundation, have often spent decades protecting open spaces, working for environmental justice and food security, advocating for renewable energy, watchdogging industries, and lobbying Congress and state legislatures. So, too, have dozens of volunteers and other organizational representatives. Academics from colleges and universities throughout the country and the world have devoted their lives to studying the environment as well as environmental politics, economics, law, and science.

The history of environmental law is also replete with the contributions of "ordinary" citizens, like Lois Gibbs and Judy Bonds, who engaged in the political process. A self-proclaimed housewife, Lois Gibbs founded the Love Canal Homeowners' Association in the late 1970s when toxic chemicals began to ooze into the basements of homes at Love Canal in New York. This group eventually persuaded the government to evacuate the entire community—over eight hundred households—in the face of serious hazardous waste contamination, and it lobbied successfully for the passage in 1980 of a national law to deal with abandoned hazardous waste sites—the Comprehensive Environmental Response, Compensation, and Liability Act of 1980 (CERCLA, the so-called Superfund law, which gave teeth to the cleanup effort in Libby, Montana). Gibbs would go on to establish the Citizens Clearinghouse for Hazardous Waste in 1981 (later renamed the Center for Health, Environment, and Justice), which is dedicated to providing technical and organizing support to individuals and communities facing toxic hazards.[53] The amazing story of Judy Bonds is part of Chapter 6. Countless others, whose stories may never be told, have accepted the challenge of tackling the environmental issues that face their neighborhoods.

There is no dearth of environmental heroes among us within our communities, public institutions, and nonprofit and private organizations. But the earth needs all of us to step up and help—and mitigating the impact of climate change is an area in which we can all lend a hand. Most scientists agree that the most pressing environmental challenge of this century is climate change. Scientific evidence for a warming planet is unequivocal. The Intergovernmental Panel on Climate Change (IPCC), a body of 1,300 scientists working under the auspices of the United Nations, has found more than a 90 percent probability that human actions over the past 250 years have led to global warming. As shown in Figure 1.1, research from international scientific institutions shows that the planet has been rapidly warming in the last few decades.[54] Global temperatures continue to rise, with ten of the warmest years in recorded meteorological history having occurred since 2002, and 2015, 2016, and 2017 being the hottest years on record. The oceans

are warming, ice sheets are shrinking, Arctic sea ice is declining, and gla-
ciers are retreating almost everywhere around the world.[55] NASA's Global
Climate Change data reveal that carbon dioxide (CO_2), a major greenhouse
gas, increased to over 406 parts per million (ppm) in 2017. For some time,
scientists have considered 350 ppm to be the threshold level for avoiding the
most devastating effects of climate change.

The consequences are unimaginable for extreme weather events, and
the evidence is clear. Those in coastal or flood-prone regions will face more
hurricanes and rising seas, as we saw in the summer of 2017 when Hurricane Harvey brought torrential rains to Texas; Hurricane Irma struck the
US Virgin Islands as a Category 5 hurricane, leaving a path of destruction
across the island and a wide swath of Florida; and Puerto Rico and the Virgin Islands were devastated by Maria, a Category 4 hurricane. If our current
level of CO_2 and other greenhouse gas emissions continues unabated, major
cities around the world will be at risk. The World Bank estimates that more
than 360 million urban residents live in coastal areas less than thirty feet
above sea level and are especially vulnerable.[56]

But a warming planet affects all of us. Those of us who live in areas that
are already hot and dry will experience hotter and dryer conditions, leading to drought, water shortages, and more wildfires. Using a climate change
model, the National Oceanic and Atmospheric Administration (NOAA)
predicted that the risk of very large fires could increase by 600 percent by
midcentury, mostly across the western United States.[57] In 2017, wildfires in
Montana, Idaho, California, Utah, Washington, and Oregon sent plumes of
smoke across the country. Montana and California, with hotter and dryer
conditions than in any recent memory, were exceptionally hard hit. Wildfires
burned for months in Montana, covering nearly half a million acres. California fires were the most destructive in state history. Climate change effects
will reach even those of us who don't live in coastal areas or in the West. For
example, diseases caused by insects are likely to increase. The incidence of
Lyme disease, a bacterial disease transmitted to humans through tick bites,
has nearly doubled in the Northeast and Upper Midwest states since 1991.[58]

Addressing climate change may help solve other environmental problems. According to *The Ecological Footprint Atlas*, the world is facing an
ecological deficit.[59] To put it simply, humanity's demands on nature now exceed the regenerative capacity of the earth. The World Wildlife Fund International's 2014 *Living Planet Report* is both grim and urgent: our footprint
now exceeds the world's ability to regenerate by about 50 percent.[60] To put
it another way, 1.5 earths would currently be required to meet humanity's
annual demands on nature. With insufficient biological materials to absorb

FIGURE 1.1: Global Land-Ocean Temperature Index

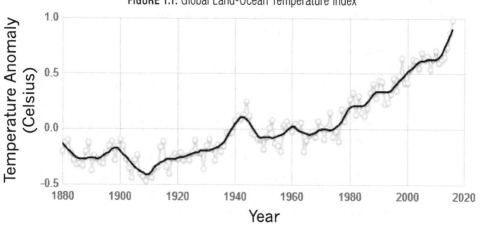

Source: Goddard Institute for Space Studies (GISS), "Global Land-Ocean Temperature Index," at NASA, Global Climate Change: Vital Signs of the Planet, "Global Temperature," https://climate.nasa.gov/vital-signs/global-temperature/.

the waste materials generated by humans or to accommodate humans' increasing demand on natural systems, our planet teeters on the brink. Critical ecosystems are collapsing, water supplies are becoming more degraded and scarcer, there is a growing loss of biodiversity, and food shortages are increasing. We are living beyond our means and, in the words of the report, risk "potentially catastrophic changes to life as we know it."[61]

As dire as the future may seem, we have the power to mitigate the worst of the consequences of climate change. Human ingenuity is creating better solar panels, wind turbines, and other sources of renewable energy. Many countries and cities have already acted: in April 2017, Germany set a new national record, generating 85 percent of its energy from renewable sources.[62] The historic Paris Agreement to cut greenhouse gas emissions across the planet was signed by 195 countries, though President Donald Trump indicated in a speech in June 2017 that the United States would withdraw from the agreement. This action, however, prompted an outpouring of increased support for addressing climate change among US citizens, corporations, states, and cities. At least ten states and major cities indicated that they would commit to the goals of the Paris climate accord independently of the federal government. The US Conference of Mayors adopted a resolution, "Supporting a Cities-Driven Plan to Reverse Climate Change," in 2017 with a goal of cities moving to 100 percent renewable energy by 2035.[63] As of August 2017, forty US cities had pledged to transition entirely to renewable

energy. Zero-net emissions of fossil fuels may yet be possible in our life-
times. We have the power to influence what our governments do, and we
have within ourselves the ability to minimize our resource consumption—
our own ecological footprint.

If history is filled with so many heroes, why is it that we have yet to
solve environmental challenges forty years after the 1970s, the first decade
of environmental activism? Why are we in danger of compromising the
planet and leaving a warming world to our children and grandchildren?
One explanation seems clear: although it took courage to create major environmental laws in the first environmental decade, it takes equal vigilance
and courage from citizens to put them into effect decades later. Time is
of the essence. The effects of climate change—including record-breaking
temperatures and extreme weather events—are escalating. The actions that
have already been taken will help, but more—much more—is needed. If the
rate of climate change is controllable, it requires "the rest of us" to be everyday environmental heroes.

In a speech to the National Council for Science and the Environment
in 2010, James Gustave Speth noted that "creating circumstances for positive change inevitably leads to political arenas, where muscular democratic
forces steered by an informed and engaged citizenry are vital."[64] That's
where we come in. That's where civic environmentalism starts. We have
the stuff of heroes in us. We can get involved and make changes that, taken
together, will yield positive results. In our sphere of influence, we can be
engaged. Wilferd A. Peterson, a twentieth-century author known for his
inspirational writings, put it well:

> Few of us will do the spectacular deeds of heroism that spread them-
> selves across the pages of our newspapers in big black headlines. But
> we can all be heroic in the little things of everyday life. We can do the
> helpful things, say the kind words, meet our difficulties with courage
> and high hearts, stand up for the right when the cost is high, keep our
> word even though it means sacrifice, be a giver instead of a destroyer.
> Often this quiet, humble heroism is the greatest heroism of all.[65]

CONCLUSION

This chapter has presented an overview of the major concepts explored in
subsequent chapters. Key concepts include the nature of industrial disasters, what constitutes a normal accident versus one caused by organizations

that have neglected safety and environmental standards, and the principles that shape high-reliability organizations. Also explored were the factors that prompt new laws and changes in policies, as government responds to industrial disasters. Finally, we looked at what makes an individual a hero or a villain, and the need for all of us to engage in environmental protection, especially as it relates to climate change.

In Chapters 2 through 5, we will take a deeper look at each of the industrial disasters briefly discussed in this chapter. Each story is unique, but they share the characteristics presented in Box 1.1. Each involves violations of multiple environmental laws, policies, or regulations, and criminal or civil charges have resulted. Each story presents a case of what might be considered environmental villainy: behavior that seems outside of what we expect from individuals and companies. But each story is also filled with courage and hope and examples of ordinary people getting involved to sound the environmental alarm or try to help their fellow citizens.

Chapter 6 offers portraits of three environmental heroes who brought leadership, courage, integrity, grit, and dedication to their efforts on behalf of the greater environmental good. Former US senator Gaylord Nelson recognized the importance of creating a social movement around Earth Day to usher in the first environmental decade, the 1970s. William Ruckelshaus served twice as the administrator of the US Environmental Protection Agency, establishing it as an important safeguard for the environment. Judy Bonds took on the issue of mountaintop removal of coal and called for a new environmental activism. She would often challenge the crowds gathered to hear her, "You are the ones you've been looking for."

As a result of the work of these heroes and others like them, we are moved to be better, kinder, more engaged environmental stewards. And so, in conclusion, Chapter 7 addresses the rest of us. We can be unsung heroes who hold organizations and the government accountable, who try to prevent or lessen the impact of industrial disasters, who work to protect the environment and our collective future. Each of us can make the world a better place as we make choices every day to address climate change, our biggest environmental challenge and one that we all share. After all, the Blue Planet is home to us all, and its story is our story.

NOTES

1. Dianne Vaughan, "The Dark Side of Organizations: Mistake, Misconduct, and Disaster," *Annual Review of Sociology* 25, no. 1 (1999): 271–305.

2. Charles Perrow, *Normal Accidents: Living with High-Risk Technologies* (New York: Basic Books, 1984; updated edition, Princeton, NJ: Princeton University Press, 1999).

3. Robert K. Merton, "The Unanticipated Consequences of Purposive Social Action," *American Sociological Review* 1, no.6 (1936): 894–904.

4. Ibid., 901.

5. Vaughan, "The Dark Side of Organizations."

6. Ibid. See also Vaughan's discussion of the normalization of deviance in *The Challenger Launch Decision: Risky Technology, Culture, and Deviance at NASA* (Chicago: University of Chicago Press, 1996), in which she argues that the culture of NASA worked against the ability of staff to voice concerns about the O-rings seating properly in low temperatures.

7. See Vaughan, *The Challenger Launch Decision*; and C. F. Larry Heimann, "Understanding the *Challenger* Disaster: Organizational Structure and the Design of Reliable Systems," *American Political Science Review* 87, no. 2 (1993): 421–435.

8. *Columbia* Accident Investigation Board, *Columbia Accident Investigation Board Report*, vol. 1 (Washington, DC: US Government Printing Office, August 2003), 25.

9. Ibid., 9.

10. Ibid., 12.

11. William R. Freudenburg and Robert Gramling, *Blowout in the Gulf: The BP Oil Spill Disaster and the Future of Energy in America* (Cambridge, MA: MIT Press, 2011), 35.

12. James Reason, *Human Error* (Cambridge: Cambridge University Press, 1990).

13. Vaughan, "The Dark Side of Organizations."

14. Quoted in Environment News Service, "Gulf Oil Spill Commission: Safety Not Sacrificed to Save Money," November 10, 2010, www.ens-newswire.com/ens /nov2010/2010-11-10-01.html.

15. Peter Schwarz and Blair Gibb, *When Good Companies Do Bad Things: Responsibility and Risk in the Age of Globalization* (New York: Wiley, 1999).

16. This is paraphrasing Norton Long, who wrote that the lifeblood of administration is power in "Power and Administration," *Public Administration Review* 9, no. 4 (1949): 257–264.

17. See Matthew L. A. Hayward and Donald C. Hambrick, "Explaining the Premiums Paid for Large Acquisitions: Evidence of CEO Hubris," *Administrative Science Quarterly* 42, no. 1 (1997): 103–127; Richard Roll, "The Hubris Hypothesis of Corporate Takeovers," *Journal of Business* 59, no. 2 (1986): 197–216; and Marie A. McKendall and John A. Wagner, "Motive, Opportunity, Choice, and Corporate Illegality," *Organization Science* 8, no. 6 (1997): 624–647.

18. Sara Ganim and Sarah Jorgensen, "Four More Charged in Flint Water Crisis," *CNN*, December 20, 2016, http://cnn.it/2h8pKLV (accessed January 31, 2017).

19. Environment News Service, "Gulf Oil Spill Commission."

20. Coral Davenport and John Schwartz, "BP Settlement in Gulf Oil Spill Is Raised to $20.8 Billion," *New York Times*, October 5, 2015.

21. Environment News Service, "Gulf Oil Spill Commission."

22. David Hammer, "BP Was More Than $40 Million over Budget for Blown-out Well, Oil Spill Hearings Show," *(New Orleans) Times Picayune*, August 26, 2010.

23. National Commission on the BP *Deepwater Horizon* Oil Spill and Off-shore Drilling, *Deep Water: The Gulf Oil Disaster and the Future of Offshore Drilling: Report to the President* (Washington, DC: US Government Printing Office, January 2011), www.gpo.gov/fdsys/pkg/GPO-OILCOMMISSION/pdf/GPO-OILCOMMISSION.pdf (accessed June 6, 2017).

24. Ibid., 84.

25. Norman G. Page et al., *Report of Investigation: Fatal Underground Mine Explosion, April 5, 2010, Upper Big Branch Mine–South, Performance Coal Company, Montcoal, Raleigh County, West Virginia, ID No. 46-08436* (Washington, DC: US Department of Labor, Mine Safety and Health Administration, Coal Mine Safety and Health, 2011), https://arlweb.msha.gov/Fatals/2010/UBB/FTL10c0331noappx.pdf.

26. Ian Urbina, "US Said to Allow Drilling Without Needed Permits," *New York Times*, May 13, 2010.

27. "Six Years Later: Remembering the Upper Big Branch Disaster" (Washington, DC: US Department of Labor, MHSA, April 5, 2016), www.msha.gov/news-media/events/2015/12/29/six-years-later-remembering-upper-big-branch-disaster (accessed May 26, 2017).

28. Todd La Porte, "High Reliability Organizations: Unlikely, Demanding, and at Risk," *Journal of Contingencies and Crisis Management* 4, no. 2 (1996): 60–71; Karl E. Weick, Kathleen Sutcliffe, and David Obstfeld, "Organizing for High Reliability: Processes of Collective Mindfulness," *Crisis Management* 3 (2008): 81–123; Todd R. La Porte and Paula M. Consolini, "Working in Practice but Not in Theory: Theoretical Challenges of 'High Reliability Organizations,'" *Journal of Public Administration Research and Theory: J-PART* 1, no. 1 (1991): 19–48; Karl Weick and Kathleen Sutcliffe, *Managing the Unexpected: Assuring High Performance in an Age of Complexity*, University of Michigan Business School Management Series (San Francisco: Jossey-Bass, 2001).

29. Karl E. Weick and Kathleen M. Sutcliffe, *Managing the Unexpected: Sustained Performance in a Complex World* (San Francisco: Jossey-Bass, 2015).

30. Ibid., 10.

31. Yuri Mishina, Bernadine J. Dykes, Emily S. Block, and Timothy Pollock, "Why 'Good' Firms Do Bad Things: The Effects of High Aspirations, High Expectations, and Prominence on the Incidence of Corporate Illegality," *Academy of Management Journal* 53, no. 4 (2010): 701–722.

32. Ibid., 716.

33. Garrett Hardin, "The Tragedy of the Commons," *Science* 162, no. 3859 (1968): 1243–1248, doi:10.1126/science.162.3859.1243.

34. Michael D. Reagan, *Regulation: The Politics of Policy* (Boston: Little, Brown and Co., 1987).

35. Thomas A. Birkland, *Lessons of Disaster: Policy Change After Catastrophic Events* (Washington, DC: Georgetown University Press, 2006); Thomas A.

Birkland, "Learning and Policy Improvement After Disaster: The Case of Aviation Security," *American Behavioral Scientist* 48, no. 3 (2004): 341–364.

36. Perrow, *Normal Accidents,* 360 (emphasis added). See also Charles Perrow, *The Next Catastrophe* (Princeton, NJ: Princeton University Press, 2007).

37. James W. Dearing and Everett M. Rogers, *Communications Concepts 6: Agenda-Setting* (Thousand Oaks, CA: Sage Publications, 1992).

38. Thomas A. Birkland, *After Disaster: Agenda Setting, Public Policy and Focusing Events* (Washington, DC: Georgetown University Press, 1997).

39. Dearing and Rogers, *Communications Concepts 6.*

40. See Lutz Erbring, Edie N. Goldenberg, and Arthur H. Miller, "Front-Page News and Real-World Cues: A New Look at Agenda-Setting by the Media," *American Journal of Political Science* 24, no. 1 (February 1980): 16–49; David A. Rochefort and Roger W. Cobb, *The Politics of Problem Definition: Shaping the Policy Agenda* (Lawrence: University Press of Kansas, 1994); see also Roger W. Cobb and Charles Elder, *Participation in American Politics: The Dynamics of Agenda-Building* (Baltimore: Johns Hopkins University Press, 1983), and the often-cited first work by Cobb and Elder, "The Politics of Agenda-Building: An Alternative Perspective for Modern Democratic Theory," *Journal of Politics* 33, no. 4 (1971): 892–915.

41. Anthony Downs, "Up and Down with Ecology: The Issue-Attention Cycle," *Public Interest* 28 (Summer 1972): 38–50, 38.

42. See, for example, Frank R. Baumgartner and Brian D. Jones, *Agendas and Instability in American Politics* (Chicago: University of Chicago Press, 1993); and Cobb and Elder *Participation in American Politics.*

43. See Cobb and Elder, "The Politics of Agenda-Building"; John Kingdon, *Agendas, Alternatives, and Public Policy,* 2nd ed. (New York: Longman, 1995); and, more recently, Thomas Birkland, "Focusing Events, Mobilization, and Agenda-Setting," *Journal of Public Policy* 18, no. 1 (1998): 53–74, and Birkland's book-length treatment on the topic, *After Disaster.*

44. Kingdon, *Agendas, Alternatives, and Public Policy,* 87.

45. Baumgartner and Jones, *Agendas and Instability.*

46. Frank R. Baumgartner, Bryan D. Jones, and Peter B. Mortensen, "Punctuated Equilibrium Theory: Explaining Stability and Change in Public Policymaking," in *Theories of the Policy Process,* 3rd ed., edited by Paul A. Sabatier and Chris M. Weible (Boulder, CO: Westview Press, 2014).

47. Denise Scheberle, "Radon and Asbestos: A Study of Agenda Setting and Causal Stories," *Policy Studies Journal* 22, no. 1 (1994): 74–86, doi:10.1111/j.1541–0072.1994.tb02181.x.

48. Andrew Bernstein, "The Philosophical Foundations of Heroism," *Capitalism Magazine,* September 25, 2004, www.capitalismmagazine.com/2004/09/the-philosophical-foundations-of-heroism/.

49. Judy Logan and Gail Evenari, "In Search of Heroes: An American Journey: The Heroism Project: High School Curriculum" (Maiden Voyage Productions, 2001), www.heroism.org/HeroismHighSample.pdf (accessed September 24, 2010).

50. Bernstein, "The Philosophical Foundations of Heroism."

51. James M. Kouzes and Barry Z. Posner, *The Leadership Challenge: How to Make Extraordinary Things Happen in Organizations*, 5th ed. (San Francisco: Jossey-Bass, 2012).

52. Norma M. Riccucci, *Unsung Heroes: Federal Execucrats Making a Difference* (Washington, DC: Georgetown University Press, 1995).

53. Center for Health, Environment, and Justice, "About CHEJ," http://chej.org /about-us/.

54. NASA, Global Climate Change: Vital Signs of the Planet, "Facts: Scientific Consensus: Earth's Climate Is Warming," https://climate.nasa.gov/scientific -consensus/ (accessed May 29, 2017).

55. For more information, see NASA, Global Climate Change: Vital Signs of the Planet (website), http://climate.nasa.gov/.

56. The World Bank, "Cities and Climate Change: An Urgent Agenda," *Urban Development Series Knowledge Papers*, vol. 10 (December 2010).

57. Caitlyn Kennedy, "Risk of Very Large Fires Could Increase Sixfold by Mid-Century in the US," NOAA Climate.gov, August 26, 2015, www.climate.gov /news-features/featured-images/risk-very-large-fires-could-increase-sixfold-mid -century-us (accessed September 15, 2017).

58. EPA, "Climate Change Indicators: Lyme Disease," 2016, www.epa.gov/climate -indicators/climate-change-indicators-lyme-disease (accessed September 15, 2017).

59. B. Ewing, S. Goldfinger, A. Oursler, A. Reed, D. Moore, and M. Wackernagel, *The Ecological Footprint Atlas 2009* (Oakland, CA: Global Footprint Network, 2009).

60. World Wildlife Fund International (WWF), *Living Planet Report 2014* (Gland, Switzerland: WWF, 2014), 9, www.worldwildlife.org/pages/living-planet-report -2014 (accessed June 3, 2016).

61. Ibid., 10.

62. Kristin Houser, "Germany Just Broke Its Own Energy Record and Generated 85% of Electricity from Renewables," World Economic Forum, May 18, 2017, www.weforum.org/agenda/2017/05/germany-just-broke-its-own-energy -record-and-generated-85-of-electricity-from-renewables-1a0ce984-9341-47ef-9ffe -4b4e17670627 (accessed May 29, 2017).

63. US Conference of Mayors, "Supporting a Cities-Driven Plan to Reverse Climate Change," 2017 Adopted Resolutions, http://legacy.usmayors.org/resolutions /85th_Conference/proposedcommittee.asp?committee=Energy (accessed September 15, 2017).

64. James Gustave Speth, "A New American Environmentalism and the New Economy," Tenth Annual John H. Chafee Memorial Lecture, National Council for Science and the Environment, Washington, DC (January 21, 2010), http://capital institute.org/wp-content/uploads/2014/08/A_New_American_Environmentalism _and_the_New_Economy_final1.pdf.

65. Wilferd Peterson. AZQuotes.com, Wind and Fly LTD, 2017, http://www .azquotes.com/quote/557574/ (accessed December 22, 2017).

CHAPTER 2

The Night of the Gas

UNION CARBIDE IN BHOPAL, INDIA

Shortly after midnight on December 3, 1984, a storage tank at a Union Carbide pesticide factory in Bhopal, India, began to leak methyl isocyanate (MIC) gas, a gas more toxic than cyanide. A runaway chemical reaction was underway in the tank—a reaction that would cause over forty tons of lethal gas to shoot into the atmosphere around the plant. The workers at the Bhopal plant felt the explosion and fled. The residents of Bhopal would not fare so well.

As the poisonous gas enveloped the densely populated city, a tragedy unparalleled in industrial history began. People began to die, especially throughout the shanty settlements that bordered the Union Carbide plant. By morning, the bodies of the victims could be found in homes, in the streets, in the countryside, and at the local hospital. In the span of a few days, 3,787 people would die in the world's worst industrial disaster, later described by the Indian government as a "holocaust" and by others simply as the "Night of the Gas."[1]

This chapter begins with an overview of the emergence of Union Carbide as a major US corporation and the Indian plant's background and operations. It then traces the disputes about who was to blame for the accident and the ensuing legal battles. The chapter concludes with a look at the legislative and legal responses in the United States, including a discussion of how Bhopal spurred the passage of a new environmental law in this country, the Emergency Planning and Community Right-to-Know Act of 1986, and the enduring legacy of Bhopal. The chapter spotlights the common characteristics of industrial disasters, the role of government in shaping environmental stories, and the part played by disasters in setting and propelling policy agendas.

BACKGROUND AND OPERATIONS OF UNION CARBIDE IN INDIA

Before the Bhopal disaster, Union Carbide had a long history as a US company, as well as a long relationship with India. After an initial operation, founded in 1898, as a manufacturer of calcium chloride in Virginia, the Union Carbide and Carbon Corporation was incorporated in 1917. (The name would be changed simply to Union Carbide in 1957.) Three years later, the company established a new subsidiary, the Carbide and Carbon Chemicals Company, and completed the first commercial ethylene plant at Clendenin, West Virginia. This established the company's chemical business. In its public relations materials, Union Carbide declared that this plant's production of synthetic organic chemicals launched the petrochemical industry.[2]

The corporation grew over the next decades as it responded to military needs in both world wars by providing a range of chemicals and products. After World War II, Union Carbide flourished when it entered the consumer market by providing plastics for consumer products as well as its Eveready batteries and antifreeze products.[3] In 1959, the company formed a highly profitable consumer products division based primarily on its plastics. By the mid-1970s, Union Carbide's agricultural products division was also a robust part of the company. The pesticides business was flourishing in part because of the company's popular pesticide, Sevin, which contained MIC. Sevin was made at Union Carbide's plant in Institute, West Virginia, where MIC was manufactured and stored. By 1984, the year of the Bhopal tragedy, Union Carbide was the thirty-fifth-largest industrial company in the United States and a well-established international corporation, with over 130 subsidiaries or affiliated companies in thirty-eight countries.[4] It was also the third-largest chemical company in the United States, marketing more than 150 chemicals.[5]

Union Carbide was one of the first US companies to expand in India, with an initial presence in that country in 1905; Union Carbide India Limited (UCIL) was established in 1934.[6] In the mid-1960s, India faced an extreme food shortage and requested food assistance from the United States.

Subsequently, the twenty thousand tons of US grain that India received daily helped to avert a major famine. In addition to food, India received agricultural supplies. India had responded to the food shortages by launching the "Green Revolution" to increase cropland and expand food production, and Union Carbide saw opportunities for its pesticide products to be produced and marketed there. By providing goods considered to be in the national interest (pesticides for India's growing agricultural needs), Union Carbide was exempted from the Indian law prohibiting foreign firms from holding a controlling interest in an Indian company.[7] Thus, Union Carbide retained majority ownership (50.9 percent) in UCIL, with the remaining shares held by the Indian government and individuals.

In 1969, UCIL established the fateful plant in Bhopal, the capital of the state of Madhya Pradesh in central India, with the blessing of the national government. The plant was originally designed by the parent company to formulate Union Carbide's carbamate pesticides, trade-named Sevin and Temik, which were common but highly toxic formulations.[8] Thus, the plant mixed and packaged pesticides with MIC imported from the Union Carbide plant at Institute, West Virginia. This would soon change.

In 1975, Union Carbide began an ambitious expansion project at the plant. The company had decided to integrate backwards: the Bhopal pesticides factory would now directly manufacture the ingredients of Sevin—MIC and alpha-naphthol—sparing Union Carbide the expense of shipping MIC overseas. In 1978, the alpha-naphthol manufacturing unit was established, and the MIC unit was built a year later, designed for an annual production capacity of five thousand metric tons. Meanwhile, this small facility on the outskirts of Bhopal had become surrounded by squatters, drawn by the roads and water lines going into the plant. The local government seemed unconcerned, however, about the shantytowns around the factory. When a local administrator, M. N. Buch, asked that the plant be moved to a new location away from the people, his pleas went unanswered, and he was promptly removed from his job.[9] With the government's blessing, the plant would remain—setting the stage for the worst industrial disaster in modern history.

LIVING AND DYING DURING THE NIGHT OF THE GAS

On the "Night of the Gas," some people died quickly or in their sleep. They may have been the lucky ones. Others woke gasping and choking, unable to breathe and with no idea how to escape the deadly gas. Many died as their lungs filled with fluid, essentially drowning; others suffered massive heart attacks. No alarms sounded from the Union Carbide plant, and city residents—even those living in the shadow of the factory—had not been forewarned about the toxic chemicals stored, used, and manufactured at the plant.[10] No public-address system advised Bhopal residents about what to do. Few knew that it might be safer to stay in their beds with a wet cloth covering their face and to position towels in doors and windows to block the gas from entering their homes. Instead, people ran choking and gasping into the streets as their eyes burned. They ran away from the Union Carbide plant, but many unknowingly followed the path of the lethal plume, which would eventually cover forty square kilometers (over fifteen square miles). Each labored breath drew in more deadly gas. The toxic cloud was dense and searing; people panicked in a frenzied effort to save their lives and the lives of their families. They blindly followed one another without knowing where they were headed, hoping to escape.[11]

The loss of life was staggering: estimates vary, but at least twenty-five hundred people died that night, and one thousand more succumbed in the week that followed. Within the next few years, roughly two thousand more deaths were directly attributed to the leak, bringing the death toll in the aftermath of the disaster to over five thousand people, according to the Bhopal Gas Tragedy Relief Department. By 2003, over fifteen thousand death claims had been processed by the Indian government in which the cause of death was attributed to exposure to the lethal gas that one night.[12] Some estimates have suggested that many more residents died from the Night of the Gas, perhaps as many as thirty thousand, but were not accounted for in government records. Cemeteries could not accommodate the massive death toll, so mass burials and cremations occurred. Many children died who were never identified.

In the aftermath of the explosion, 150,000 people sought treatment at hospitals and clinics, overwhelming medical facilities. In the years that followed, an estimated 200,000 survivors developed symptoms from exposure to the gas, and by the official reckoning, a staggering 578,000 people were affected in some way by the catastrophe.[13] The horrifying catalog of debilitating symptoms caused from exposure to the deadly poison included impaired vision or blindness, respiratory illness, neurological issues, depression, reproductive ailments, and birth defects.

The tragedy disproportionately affected the poorest of the poor who lived around the plant—the very ones with the fewest resources to cover the health and economic costs of the catastrophe. Although the chief minister of the Madhya Pradesh state promised compensation for the victims ($100 for families with a hospitalized family member and $500 for families that had suffered a death), it was too little and too late.[14]

Photo 2.1: Two men carry children blinded by the Union Carbide chemical pesticide leak to a hospital. Thousands were killed or seriously injured in the gas leak. *AP Photo/Sondeep Shankar.*

In addition to the loss of human life and the profound health consequences for those who survived the initial disaster, exposure to the gas ravaged the animal population, both domestic and wild. Over 2,200 livestock died that night, along with countless numbers of domestic animals and wildlife.[15] The economic consequences were also severe: in the area of the disaster, businesses were shuttered and street markets came to a halt, since many of those afflicted had worked as street vendors.

As complete chaos enveloped the city during the Night of the Gas, the local government could not deal with the unfolding tragedy. In what was described in eyewitness accounts as a "total system breakdown," people were left to cope on their own. According to the state government of Madhya Pradesh, thirty-six of the fifty-six wards in Bhopal were affected by the disaster, representing half of the population of Bhopal.[16] When the one hospital serving the area was overwhelmed, illegitimate doctors, pharmacies, and lawyers were quick to profit from the sufferings of the poor.

Compounding the desperate situations of the victims, the government created more chaos in the gas-affected communities near the plant. Instead of delivering food and drinking water at the doorsteps of people suffering from exposure to the MIC, the administration asked victims to stand in line, even though most were incapable of doing so.[17] Unscrupulous people grabbed relief rations intended to help the gas victims. Ultimately, the dev-astation was of such a magnitude that it prompted the establishment of a new government agency, the Department of Bhopal Gas Tragedy Relief and Rehabilitation, and the extended involvement of the state government of Madhya Pradesh. To this day, the Indian government maintains a relief agency to serve the needs of the survivors and children of Bhopal.

The explosion marked the end of the Bhopal facility. Abandoned, most of its toxic contents remained at the facility, leaching into the soil, the groundwater, and the surface water surrounding the plant. The poor con-tinued to take their drinking water from these contaminated supplies for years, and children played on contaminated soil in and around the plant, thus becoming what is known as the second generation of Bhopal.

THE ENVIRONMENTAL CONSEQUENCES
OF THE BHOPAL DISASTER

If the human cost of the Night of the Gas was enormous, the environmen-tal consequences were also severe. In the days following the tragedy, many short-term environmental consequences were evident. In addition to the loss of at least two thousand animals, concern that the rivers and lakes were pol-luted prevented people from fishing near the site. Crop growth was affected, as were trees near the site. Bhopal would overcome these short-term effects, but the long-term environmental consequences proved much more difficult.

It would be nice to be able to say that the environmental story is over. It is not. After the gas leak, the factory was closed, never to reopen, but Union Carbide made little effort to remove its stocks of lethal chemicals. The toxic remains of the pesticide factory lasted for decades, contaminating soils and groundwater—the primary source of drinking water for fifteen commu-nities around the factory in Bhopal.[18] Union Carbide maintained that no contaminated soil or groundwater was directly caused by the MIC gas leak, an assertion confirmed by some Indian government authorities.[19] However, a report released in 2009 by the UK-based Bhopal Medical Appeal and a medical clinic, the Sambhavna Clinic, suggested that the water contami-nation was worse than before the Night of the Gas some twenty-five years earlier. The report noted that, "not surprisingly, the populations in the areas surveyed have high rates of birth defects, rapidly rising cancer rates, neuro-logical damage, chaotic menstrual cycles, and mental illness."[20]

Some scientists suggested that the facility had routinely neglected en-vironmental standards for years, dumping highly toxic wastes inside its sixteen-acre factory site and into ponds designed to store liquid waste from the plant.[21] A former employee at the UCIL plant stated in his affidavit that between 1969 and 1984 huge quantities of pesticides, solvents, and toxic waste were routinely dumped at the site.[22] While Union Carbide pointed out that three large evaporation ponds, with plastic liners, covering 35 acres some 350 yards north of the facility legally received thousands of gallons of liquid, hazardous substances between 1969 and 1984, it remained silent regarding any illicit dumping.[23]

In 1989, amid increasing public charges that the factory was causing illness, Union Carbide privately conducted an investigation of possible contamination. The testing revealed that the soil and water inside the factory were massively contaminated. The majority of Union Carbide's samples contained alpha-naphthol or Sevin in quantities "far more than permitted for on-land disposal."[24] The testing showed that these poisons were sufficiently concentrated in the samples to cause 100 percent fish mortality. However, Union Carbide neither informed the public of the test results nor let them know of the potential danger to local water supplies.[25]

Over time, the monsoons battered the shuttered facility as well as the ponds, which overflowed with a toxic chemical soup. Wells that tapped into this aquifer were poisoned. Environmental and citizen groups continued to focus attention on the lethal chemicals in rotting drums and sacks at the abandoned site. In 1999, a Greenpeace report revealed what Union Carbide would not: the factory site, surrounding land, and groundwater were severely contaminated.[26] Mercury levels were found in some places on-site that were six million times higher than background level. Drinking water wells near the factory remain heavily polluted to this day with chemicals known to produce cancers and genetic defects. Greenpeace scientists warned policymakers that it was "essential that steps are taken to reduce and, as far as possible, eliminate further exposure of communities surrounding the contaminated site to hazardous chemicals. Contaminated wastes and soils must be safely collected and securely contained, until such time as they can be effectively treated. Such treatment must entail the complete removal and isolation of toxic heavy metals from the materials, and complete destruction of all hazardous organic constituents."[27] However, no substantive efforts to clean up the site occurred after the investigation, and the state government only began to address cleanup in 2012.

No laws forced Union Carbide to clean up its plant, in part because of the settlement with the Indian government, but also because Union Carbide no longer existed as a corporate entity. In 2001, Dow Chemical acquired Union Carbide as a wholly owned subsidiary. The US law governing the cleanup of abandoned waste sites, our Superfund law, does not apply overseas.

Dow has resisted calls to remove toxic materials on-site or to clean contaminated groundwater. Dow has repeatedly refused to accept any liability for Bhopal, despite protests and legal challenges.[28] In 2010, India's attorney general, Goolam Vahanvati, asked the country's Supreme Court to force Dow Chemical to pay $1.1 billion to compensate Bhopal victims. On December 3, 2011, the twenty-seventh anniversary of the Night of the Gas, massive protests against Dow led to arrests. Dow's involvement as a sponsor of the 2012 Summer Olympic Games in London prompted protests and a formal demand by India's Olympic body to remove Dow as a sponsor. In 2015, Dow claimed that it had no obligation to respond to a summons to appear in court regarding the Bhopal gas leak, stating:

> It is important to recognize that Dow never owned or operated the Bhopal facility and any efforts to directly involve Dow in legal proceedings in India concerning the 1984 Bhopal tragedy are misguided and without merit. Dow has no liability for Bhopal and any attempts to attach the company to the criminal matter are highly inappropriate; criminal liability cannot be transferred from one entity to another under any circumstance. Finally, and importantly, Dow is not subject to criminal jurisdiction in India, just as UCC is not.[29]

Dow merged with DuPont in 2017, prompting renewed concerns among the victims of the "second disaster of Bhopal" that the environment will not be restored, nor will health issues be addressed. Believing that the polluter should pay, most advocates for the victims hold Union Carbide, UCIL, Dow, and now Dow-DuPont responsible for the site cleanup and for the ongoing medical needs of the victims.

However, the Madhya Pradesh government has been slow to respond as well. It would take until 2009 before a rudimentary piping system was provided by the city government, and until 2011 before piped drinking water was made available to residents around the factory.[30] In 2012, the Indian government was still wrangling with decisions about safely disposing of the 390 metric tons of waste at the defunct site.[31] The state conducted a trial incineration of just 10 metric tons of waste in 2015, but the location of the incinerator was challenged by the Bhopal Group for Action and Information over environmental concerns.[32] Further incineration of toxic waste at the abandoned site will be decided by the Indian Supreme Court. Meanwhile, the country's National Institute for Research in Environmental Health continues to study the legacy of MIC exposure, speculating that a third generation might be affected.[33]

If a silver lining can be found in this environmental and public health nightmare, it is that the Night of the Gas was a triggering event for the Indian government's passage of the Environment Protection Act of 1986 and the creation of the Ministry of Environment and Forests. Under the new law, industrial plans for the country must be reviewed for their potential environmental and human health risks. The Bhopal disaster also triggered new policies in the United States, as we discuss later in the chapter.

DISASTER FORETOLD IN BHOPAL

The Bhopal disaster was by no means a "normal" accident. In this section, we take a closer look at the poor decision-making and errors in judgment on the part of both Union Carbide and the Indian government that led to the Night of the Gas.

A History of Poor Decision-Making by Union Carbide

As described in Chapter 1, industrial disasters that are not "normal" accidents have several features in common. One is an atrophy of vigilance—a complacency on the part of management that results in poor decision-making. Another is patent disregard for environmental or safety standards, usually because of an overly myopic focus on the bottom line. Both the actions and inaction of Union Carbide and UCIL leading up to the Bhopal disaster display these two characteristics.

Without a doubt, producing pesticides of the toxicity of Sevin or Temik was a hazardous undertaking. However, the years leading up to the Night of the Gas suggest a series of errors in judgment about safety procedures, plant design, and MIC production. Union Carbide and UCIL made a number of decisions that magnified the devastation and loss of life that occurred on December 3, 1984. The first decision by Union Carbide that would prove fatal was linked to its plan to expand the facility so that Sevin could be fully manufactured on-site, rather than wait to receive MIC from Union Carbide's West Virginia plant. In 1979, the company built a substantial storage capacity for the MIC gas. Instead of designing the MIC plant for nominal storage (that is, to store just the amount of gas needed for downstream production), Union Carbide chose to design the facility for large-scale storage. This decision was made by the parent company even though its subsidiary (UCIL) felt that nominal storage was inherently safer.[34]

The sworn affidavit of Edward Munoz, general manager of the agricultural products division at UCIL, bears witness to the towering presence of the Union Carbide Corporation (UCC). During fact-finding for the causes of the tragedy, Munoz testified that the parent company made all major design decisions for the new plant at Bhopal, and that the engineering department at Union Carbide in West Virginia had primary responsibility for plant construction, conceptual development, and engineering.

Munoz noted that UCIL believed that only "token storage" of MIC was necessary, and that it preferred that the deadly poison be held in "small, individual containers based both on economic and safety considerations."[35] However, Union Carbide decided to build three large tanks, each one of them forty feet by eight feet in diameter. Storing MIC in fifteen-thousand-gallon tanks was not standard business practice at the time. Bayer in Germany, Mitsubishi in Japan, and DuPont all used MIC in closed loop processes, which required little or no storage of MIC.[36] On that fateful night, nearly forty tons of MIC was released. If the storage of MIC had been more closely aligned with standard operating procedures, it is very likely that fewer lives would have been lost.

Without question, vigilance had atrophied at the plant, which had a history of taking little or no heed of failures in the system. No fewer than five chemical accidents occurred at the Bhopal plant between 1981 and 1984. On December 25, 1981, a faulty valve released phosgene gas, and a factory worker, Ashraf Khan, was killed. Less than two weeks later, twenty-five workers were hospitalized as a result of another leak at the plant. And in February 1982, a MIC leak affected eighteen workers. In April 1982, UCIL expressed concerns to Union Carbide about the ongoing issues with leaks at the plant, especially in the MIC production unit. Union Carbide responded by sending US experts to the UCIL plant to conduct an audit in May 1982. The team identified a host of issues, including thirty leaking valves, nearly half of which were located in the MIC and phosgene units.[37] In their report, the team concluded that leaking valves were "fairly common" and that "valve leakage would appear to continue to be a situation that requires continuing attention and prompt correction."[38]

Nevertheless, on October 5, 1982, hundreds of people living close to the plant were hospitalized when yet another leak occurred—a potent combination of MIC, hydrochloric acid, and chloroform. This leak prompted one union representing UCIL workers to print hundreds of brochures to distribute in the community warning residents to beware of fatal accidents and informing them that safety measures at the Bhopal plant were inadequate and endangered the lives of thousands of workers and citizens.[39]

At this point, leaks had become numerous enough that UCIL took action. But instead of rectifying the production problems that were causing these accidents, UCIL decided, in September 1982, to disconnect the plant alarm from the public siren system. This would allow only the workers to hear the warning that a leak had occurred and prevent what UCIL described as "undue panic" in neighborhoods surrounding the plant.[40]

The safety audit of the Bhopal plant in 1982 certainly would have been a wake-up call to a corporation that was concerned about worker and community safety. However, when a similar safety audit performed in September 1984 at Carbide's plant in Institute, West Virginia, noted that "runaway reactions in MIC storage tanks" were a major concern, corporate officials did not share this information with UCIL.[41] This was just three months before the explosion in Bhopal.

Perhaps one reason for Union Carbide's decisions and inaction in India was that things had not turned out as planned for the company. The Green Revolution had not provided windfall profits for UCIL or Union Carbide. There had been competition from makers of a safer insecticide made with pyrethroids, which had become the preferred insecticide for cotton. In an internal memo dated February 24, 1984, UCIL and Union Carbide Eastern (the Hong Kong unit of the multinational corporation) noted that the Bhopal plant had lost $7.5 million to date, and that "the future of the existing business does not look any brighter than the poor financial performance obtained to date."[42] As a consequence, the UCIL board was "less than enthusiastic" about additional investments in the agricultural chemistry enterprise.

Union Carbide officials lamented the steady stream of losses at the Bhopal plant and commissioned a feasibility study for relocating the plant to either Indonesia or Brazil. While waiting for word from headquarters on whether to close the plant, change business strategies, change locations, or find a buyer, UCIL engaged in draconian cost-cutting measures. UCIL also shut down the MIC production unit for six weeks (but continued storing MIC) because of an oversupply of pesticides. Among the most egregious cost-cutting measures was the decision in June 1984 to shut down the refrigeration unit.[43] This move violated established operating procedures: refrigeration was critically needed to cool stored MIC in order to prevent its vaporizing or reacting. Investigations after the disaster revealed that many systems at the plant were insufficient for such large quantities of lethal chemicals. For example, the vent gas scrubber was overwhelmed during the accident as MIC and its reaction products flowed through the scrubber at more than two hundred times its capacity.[44] The scrubber had been turned off completely just two months before in an effort to save money. The flare tower was out of service as well. And when the reaction began on December 2, 1984, escaping gas could not be directed to the tower because a corroded pipe had never been replaced.

Even if the refrigeration unit, scrubber, and flare tower had been operating properly, however, the runaway reaction, because of the quantity of MIC in the tanks, most likely could not have been stopped. Reports on the volume of MIC in tank 610 (where the reaction occurred) vary from 11,290 gallons to 13,000 gallons. Either figure, however, is well above what Union Carbide's own technical manual recommends. A second tank, tank 611, contained over 5,000 gallons, and the third tank, tank 619, which was supposed to be kept empty in case of emergencies, held unknown quantities of contaminated MIC. Either tank 611 or tank 619, if empty, might have been used as a surge tank to help gain control over the runaway reaction.

When the accident occurred, another safety protocol, the water shroud, which envelops leaking gas and prevents it from leaving the plant, was undersized for the amount of MIC onsite. The water shroud reached only 40 feet in height—the reaction would spew gas 150 feet into the air. The safety inspection by Union Carbide in 1982 recommended a larger water spray system, but neither UCIL nor Union Carbide followed through with installation.

Corroded, broken, or malfunctioning gauges and safety valves were not replaced. Broken gauges made it hard for operators at the MIC tanks to understand what was happening. The pressure indicator for the MIC tanks had been malfunctioning for more than a year before the accident. Broken and leaking valves posed a huge safety risk for a plant producing toxic and lethal products, but in yet another nod to saving money, no gas detectors capable of sensing and locating leaks were installed at the plant.

That was not the only maintenance issue. UCIL and Union Carbide had made cuts in human resources. Manpower had been severely cut to save on expenses, and the workers who were left were often placed in positions with little or no training. In the MIC facility, the production crew had been cut from twelve to six and the maintenance crew from six to two, and the supervisory position on the second and third shifts had been eliminated.[45] Work conditions were poor, and turnover was high. As a result, few workers were trained in the safety and health risks of handling MIC. Astonishingly, operating procedures were written in English, even though many workers at the plant only spoke Hindi.

Union Carbide's own 1976 safety manual lists MIC as hazardous by all means of contact and poisonous by inhalation or contact.[46] Union Carbide clearly recognized the extreme danger of MIC exposure, warning in the manual that MIC was so volatile that stringent safety precautions were to be taken in handling it. The company's manual also warned that various contaminants, including water, could trigger "runaway reactions" and "hazards from the standpoint of ignition . . . and toxicity."[47] The "threshold limit value"—the maximum amount of airborne chemical substance deemed safe for full-time workers—for MIC is 0.02 parts per million for worker exposure in an eight-hour period.[48] This limit is five times more stringent than the standard for phosgene, one of the three deadly gases used as weapons in World War I (the other two being chlorine and mustard gas).

So, on that tragic night, systems were in disrepair, the MIC storage tanks were filled beyond recommended levels, no supervisor was on duty at the MIC facility, and fail-safe systems such as the water spray and refrigeration units were under-designed or simply turned off. It was in this context that the small crew began to perform what they thought was routine maintenance on the MIC production unit. They filled the lines of four process filters with water, and the water entered tank 610 through a jumper line and closed isolation valve. The exothermic reaction that followed could not be contained in part because the refrigeration unit had been turned off—a critical safety precaution because it makes the chemical much less reactive and slows down the reaction that is underway. Additionally, the MIC in tank 610 contained thirty-two times more chloroform—used as a solvent in the MIC manufacturing process—than normal. When a tragedy of incomprehensible proportions ensued, no alarm sounded—because of UCIL's decision to delink the plant and public warning systems.

Poor Decision-Making by the Indian Government

Poor decisions by the Indian government also contributed to the Bhopal disaster. One of the first was the government's response to the critical food shortages of the 1960s, which allowed Union Carbide to assume a prominent role in assisting the country in food production. Although this action might be otherwise laudable, it created a special opportunity for Union Carbide and allowed Sevin to be used and distributed throughout the country. The government also required UCIL to pay Union Carbide in US dollars for imported Sevin, a decision that ultimately pressured UCIL into producing Sevin locally.[49] This is an important point, as Bhopal was the only plant outside of the United States that formulated Sevin. Absent the government's push for developing factories in Bhopal, the factory might have remained simply a formulation plant.

Another poor government decision was allowing Union Carbide to build the MIC unit within its existing plant. While environmental and public safety laws were very limited at the time in India, the local government had acted to establish zoning laws to protect the public from exposure to hazardous materials. The plant did not conform to existing regulations.

The 1975 Bhopal Development Plan required that obnoxious and hazardous industries be located in the northeast end of the city—downwind and away from heavily populated areas. Though Union Carbide's request to build the MIC production facility in the existing plant was initially denied by local authorities in Bhopal, the central government waived this requirement in its effort to support emerging economic development.[50] In doing so, the government opted to let a hazardous operation (the storage of lethal gas and subsequent production of pesticides) occur very close to large concentrations of people. The danger was compounded by allowing the poor to create squatter settlements around the plant. Had the right-thinking requirements of the plan been upheld, the loss of life would have been reduced dramatically.

The government then compounded these errors further by ignoring the pleas of a journalist hoping to save Bhopal. Those on the audit team in 1982 were not the only ones concerned about ongoing safety issues at the plant. Raajkumar Keswani, writing for the *Rapat Weekly,* a small paper in Bhopal, wrote a series of articles warning about the dangers at the Union Carbide plant.[51] In September 1982, Keswani published his first article, titled "Save, Please Save This City."[52] He chronicled the incidents of gas leaks at Union Carbide and the dismal safety arrangements in the factory as reported by his friends working at the plant. In prophetic prose, he warned that Bhopal was "sitting at the edge of a volcano" that could spew deadly gas and kill the population. However, the *Rapat Weekly* was a small paper with fewer than three thousand readers. After no significant public outcry came from the series, Keswani decided to make his plea directly to the chief minister of Madhya Pradesh, Arjun Singh. His letter of October 15, 1982, read, in part:

> I know for a fact that the city of Bhopal is in danger, save it. . . . From the fear that I will be labeled a yellow journalist, should I be blind [to these dangers from the Union Carbide plant]? No, I will not give up, I will fight with firm determination—I will not let this city turn into Hitler's gas chamber. . . . Please take a look and see what death has in store for the living in Bhopal.[53]

One might hope that such a letter would get some attention from the minister's office, especially in light of yet another accidental release at the plant earlier that same month. But nothing came of Keswani's letter to Chief Minister Singh, or of any of his subsequent newspaper articles, the last one published just six months before the disaster. One reason might be the close relationship between state and local officials and UCIL, which

Keswani documented in his news articles. Between 1982 and 1984, Keswani published a list of government officials who benefited from the largesse of Union Carbide, including those who were frequent guests at the company's lavish guest accommodations (including, by the way, Chief Minister Arjun Singh). Keswani also noted that relatives of politicians and high-ranking officials were hired by the company, often at salaries much higher than those of workers at the plant. Not surprisingly, when Keswani brought his newspaper series to the state assembly, his fears about the Union Carbide plant were dismissed. Tara Singh Viyogi, then the labor minister, assured the assembly that he had personally seen the "foolproof safety arrangements in the Union Carbide plant."[54]

Though Keswani's coverage of Union Carbide was eventually printed in larger news venues, no substantive government action was taken to protect citizens from lethal accidental releases of gas or to force Union Carbide to reconnect the nonexistent public warning system. Keswani and his family would themselves become victims of the Night of the Gas, but were able to escape with their lives. Though he considered his work a failure because he was not able to persuade government officials to act to prevent the deadly release of MIC, others thought differently. Keswani received the prestigious B. D. Goenka Award for excellence in journalism in March 1985 for his coverage of Union Carbide.

Keswani was not the only heroic voice raising concerns about the operations of the UCIL plant. In March 1983, a lawyer in Bhopal, Shahnawaz Khan, served notice on UCIL that the plant posed a serious health risk to workers and citizens living near the plant. In his "citizen's letter to UCIL," he predicted the events that came to characterize the accident. His letter warned that "there is always the danger of an untoward accident. The lives of 50,000 people are in danger and the specter of death looms over them."[55] He went on to state that UCIL had fifteen days to stop the use of poisonous gases or legal action would ensue. UCIL denied the truth of the allegations and responded by continuing to pursue cost-cutting measures at the plant.

LEGAL BATTLES AFTER BHOPAL

Not surprisingly, a disaster of this magnitude led to protracted legal and political battles over responsibility for victim compensation and restitution. Critics would argue that the legal system moved at a glacial pace and awarded miserly compensation to victims and their families, thereby compounding their suffering. Figure 2.1 provides a timeline of key events before and after the Night of the Gas.

As the extent of the tragedy became clear in news reports across the globe, it appeared at first that the legal system would move rapidly to assist the victims. American tort lawyers arrived in Bhopal within days after the explosion, seeing an opportunity to gain hefty legal fees by representing the Bhopal residents affected by the MIC leak against Union Carbide.[56] The first lawsuit against Union Carbide was filed on behalf of thousands of victims on December 7, 1984, in a US federal district court, followed by 145 additional lawsuits.[57] The suits were subsequently joined into one action in the US District Court for the Southern District of New York in February 1985.

However, dreams of huge settlements (with correspondingly large retainers for attorneys) were short-lived. On March 29, 1985, the Indian Parliament dashed any hopes of massive tort litigation by passing the Bhopal Gas Leak Disaster (Processing of Claims) Act.[58] This law gave the state exclusive representation for all claims arising from the disaster, under the doctrine of *parens patriae* (father to the people).[59] It also provided that any compensation would be delivered by an agency of the state, under protocols established by the state's review board.

Thus, the fate of the victims now rested in the hands of the Indian government. In a regrettable testament to the state of its own laws protecting citizens from hazardous chemical accidents, India did not seek redress in its own courts but opted instead to pursue claims against Union Carbide in the United States. In 1985, India set forth a $3.3 billion claim against Union Carbide in the Southern District Court that superseded the similar claims in the 145 lawsuits, involving over 200,000 plaintiffs, that had already been filed.[60]

India's attempts to transfer legal proceedings to the United States proved difficult. The government's class action case on behalf of Bhopal's victims never went to trial. Legal skirmishes between Union Carbide and India over appropriate jurisdiction were ultimately won by Union Carbide. As a result, cases would not be heard by US federal courts. Four years later, in 1989, the Indian government settled out of court for $470 million, which ultimately amounted to less than $500 per person harmed by the accident.[61] The government's decision to settle effectively ended all outstanding civil claims against Union Carbide resulting from the gas leak, and it also ended official legal inquiry into the facts surrounding the disaster. The survivors, however, felt no such closure and continued to press for a just compensation for their injuries and loss of family members. They also demanded that criminal trials be held to bring the perpetrators of the disaster to justice.

Though charged with culpable homicide, no criminal cases have gone to trial against Union Carbide Corporation or its CEO, Warren Anderson,

FIGURE 2.1: TIMELINE OF THE "NIGHT OF THE GAS" EXPLOSION IN BHOPAL

1934: Union Carbide India Limited (UCIL) is established as a subsidiary to Union Carbide Corporation (UCC).

1968: The government of India approves UCIL plans to build a pesticides plant in Bhopal; the plant begins operating a year later.

1979: The Indian government approves Union Carbide's request to expand storage capacity for methyl isocyanate (MIC) gas.

1982: Twenty-five workers at the plant are hospitalized after exposure to gas leaks. An audit by Union Carbide reveals many safety concerns requiring correction. The journalist Raajkumar Keswani publishes his first article about the dangers at the Bhopal plant.

1983: The lawyer Shahnawaz Khan writes a citizen's letter to UCIL threatening legal action if the dangers at the plant are not corrected.

February 1984: UCIL cuts costs, including reducing the MIC staff by half and shutting down the refrigeration unit.

December 1984: Approximately forty tons of MIC gas leak from the UCIL plant. The estimated loss of life that night is over three thousand, with thousands more injured. Thousands of animals perish as well. Warren Anderson, CEO of UCC, comes to Bhopal on December 5, but leaves the following day to avoid arrest. Congress holds hearings on the Bhopal tragedy in Institute, West Virginia.

1985: The Indian Parliament passes the Bhopal Gas Leak Disaster (Processing of Claims) Act. UCC closes the Bhopal plant. UCC's plant at Institute, West Virginia, releases toxic gas, sending 135 people to the hospital.

1989: The final settlement of $470 million paid by UCC ends civil litigation against the company. It will take until 2004 for the bulk of the settlement ($357 million) to be distributed to victims.

1986: Congress passes the Emergency Planning and Community Right-to-Know Act (EPCRA).

1994: UCC sells its share of UCIL to an Indian company.

1998: The state of Madhya Pradesh assumes control of the plant site, but little remediation is done.

2001: Dow Chemical Company acquires UCC, but refuses to clean up the Bhopal plant.

2010: Seven UCIL executives are convicted of negligence and sentenced to two years in prison.

2016: The last case brought in US courts by Indian citizens seeking damages from UCC and Dow is dismissed.

2017: Over three hundred metric tons of hazardous materials remain at the site. Dow, DuPont complete merger.

despite the fact that he and eight other individuals (the eight executives at UCIL) and three corporations (UCIL, UCC, and UCE, a wholly owned subsidiary of UCC based in Hong Kong) were accused of criminal conduct just hours after the accident. Bhopal local authorities placed Warren Anderson under house arrest when he arrived in Bhopal on December 4, 1984, to inspect the damage. But he posted bail and was escorted out of the country in the chief minister's plane.

Anderson never stood in a courtroom to face charges. Shortly after his return to the United States, Anderson said that he would consider returning to India to stand trial on charges of crimes connected with the leak—including "criminal conspiracy," which under Indian law carries a maximum penalty of death—but he never returned to India.[62] Neither Warren Anderson nor any other Union Carbide official has returned to India in the decades following the tragedy to face trial. For years the Indian government requested that Anderson be extradited, and it issued a formal request in 2003. The US government never agreed to extradite him, and Anderson died in 2014, at the age of ninety-two.

Anderson insisted to the American press that neither he nor his company was responsible for the tragedy, but at the same time he claimed that "Union Carbide has a moral responsibility in this matter, and we are not ducking it."[63] In 2005, the Bhopal court issued a summons for Dow Chemical to attend the proceedings and explain why it should not produce its fully owned subsidiary, the proclaimed absconder Union Carbide, in court. Dow's subsidiary in India, Dow Chemical India Private Limited, successfully applied for the summons to be stayed.[64]

Finally, more than twenty-five years after the explosion, in a move dubbed "too little, too late," eight former executives of the company's Indian subsidiary UCIL were convicted—not of culpable homicide, but of negligence. The seven surviving defendants (one of the executives had died in the intervening years), including then–UCIL chairman Keshub Mahindra, were sentenced by a Bhopal court on June 7, 2010, to two years in prison and fined 100,000 rupees, or $2,100.[65] In a move that intensely angered the victims of Bhopal, they were granted bail immediately.

The Indian Supreme Court agreed to reopen the Bhopal case in August 2010, after the Central Bureau of Investigation (CBI) asked the court, through a curative petition, to review its own decision.[66] In 1996, the Supreme Court had diluted the charges against the Union Carbide executives from culpable homicide to criminal negligence. In its petition, the CBI said, "The men behind one of the world's biggest industrial catastrophes should not walk away with a minimal punishment of two years despite ample evidence to show the commission of an offense of homicide."[67] The Supreme Court rejected the CBI plea a year later, stating that its 1996 judgment was "never a fetter for the CBI or Madhya Pradesh government to seek enhancement of the charges."[68] The sentencing of the UCIL employees renewed calls for the extradition of Warren Anderson, even though the US State Department had declared Bhopal a "closed case" in August 2010.[69] Anderson died in obscurity, still labeled a fugitive and "absconder" by the Indian government.

CONGRESS RESPONDS TO BHOPAL

As discussed in Chapter 1, cataclysmic events often trigger a swift examination of existing laws and regulations and the creation of new ones. In the United States, the enormity of the Bhopal tragedy provided just such a "focusing event." Within two weeks after the tragedy, four subcommittees in the US House of Representatives had taken up the matter, with an eye toward preventing another such disaster, whether on US soil or abroad.[70] Congressional attention to the Bhopal tragedy would influence the reauthorization of the Comprehensive Environmental Response, Compensation, and Liability Act of 1980 (the Superfund law) and shape the way hazardous air pollutants were identified and regulated in the 1990 amendments to the Clean Air Act.

Representative James J. Florio (D-NJ), chair of the Energy and Commerce Subcommittee on Commerce, Transportation, and Tourism, pressed for stronger regulation of toxic chemicals under CERCLA and the national law dealing with hazardous waste, the Resource Conservation and Recovery Act (RCRA). In turn, Representative Henry A. Waxman (D-CA), who chaired the Energy Subcommittee on Health and the Environment, introduced a bill designed to force the EPA to better regulate hazardous air pollutants under the Clean Air Act. The two subcommittees held joint hearings on December 14, 1984, in Institute, West Virginia. As mentioned earlier, Institute was home to not just any Union Carbide plant—it was the only US facility producing MIC. In an ironic twist of history, this proved to be the perfect forum for advancing an environmental and public health agenda. Less than a year later, on August 11, 1985, the Union Carbide plant in Institute had an accidental release of a toxic gas (aldicarb oxyme) that sent 135 people to the hospital.

Waxman sought to strengthen the EPA's regulation of air toxins under Section 313 of the Clean Air Act, which required the agency to set National Emissions Standards for Hazardous Air Pollutants (NESHAP). Because these were health-based standards, however, the EPA would first have to ascertain the health risk posed by the toxic air pollutant and then set levels for the emission of that pollutant that were sufficiently protective of human health. Setting health-based standards had proven to be a daunting task for the EPA. Because the EPA was not required to regulate hazardous air pollutants until it judged them to be hazardous, based on evidence it had gathered, the onus was on the agency to understand all toxic air pollutants and also to assess the threshold level of public safety for each toxin.

Similarly, Title I of the Clean Air Act required the EPA to set National Ambient Air Quality Standards (NAAQSs) for ubiquitous pollutants—those commonly experienced across the country. However, only six pollutants were regulated under Title I, and these had been identified by Congress.[71] Every revision of the health-based standard under the main title of the law had brought a host of lawsuits. Environmental groups and health groups challenged the standard as too weak, while industrial organizations, the auto industry, and mining associations challenged the standard as too high.

If defending the NAAQS standard in court consumed large amounts of the EPA's time and energy, identifying, researching, and establishing appropriate NESHAP standards for particular hazardous air emissions proved even more challenging. At the time of the Waxman-Florio hearing, the EPA had listed only five pollutants as hazardous, and MIC was not one of them. Setting the stage for a new direction for the Clean Air Act, Waxman complained: "EPA has taken the position that until they decide to regulate a chemical, they are not going to declare it as hazardous. . . . They've been chasing their tail around for 14 years now."[72]

Florio's subcommittee focused on the recently amended RCRA and the upcoming revisions to CERCLA. One emphasis of RCRA, which regulates the treatment, storage, and disposal of hazardous waste by companies and organizations, is on the export of hazardous waste. RCRA reauthorization in 1984 prohibited exports of hazardous waste unless the EPA was notified that the receiving country had consented to accept it. This represented a legislative response to an Executive Order issued by President Jimmy Carter in the closing days of his administration (and subsequently revoked by President Ronald Reagan) that required the US government to notify foreign countries receiving a wide range of substances banned or controlled under US law, including hazardous substances, pesticides, and chemicals. It also set up procedures for banning the export of substances judged to be extremely hazardous. This provision potentially would protect foreign countries—at least with respect to waste that contained hazardous substances.

However, CERCLA's reauthorization received the most attention. Releases of hazardous substances are the central focus of the law, which gave the EPA two major responsibilities relating to hazardous materials: to respond in emergency situations to accidental releases of hazardous substances in concert with state and local authorities, and to identify the locations of hazardous substances and then assess the potential risks. If those sites (most likely abandoned) posed a risk to human or ecological health, the agency was charged with cleaning them up. The law operated under the "polluter pays" principle. This contentious Superfund authority placed strict, joint, and several liabilities on potentially responsible parties to pay for cleanup, including the companies that manufactured hazardous substances. An accident like Bhopal in the US would certainly trigger the emergency response provisions in CERCLA, but releases of hazardous substances into the environment would activate the cleanup provision as well.

On December 12, 1984, the Subcommittee on Asian and Pacific Affairs of the House Foreign Affairs Committee, chaired by Stephen J. Solarz (D-NY), held hearings to consider what requirements should be placed on US multinational corporations. Representative Solarz intended to introduce a bill that would require US companies or the government to inform foreign governments of potential health, safety, and environmental problems posed by US plants operating on foreign soil. Companies would also be required to inform foreign governments of US regulatory standards that applied to their operations in the United States. However, this bill was not likely to be supported by the Reagan administration. In testimony before the subcommittee, Robert A. Peck, deputy assistant secretary of state, reiterated that overseas manufacturing operations were usually subject to the law of the host country, not US law: "We do not generally apply US environmental and industrial safety laws to activities of multinational enterprises in other countries."[73]

On the same day, the House Education and Labor Subcommittee on Health and Safety held its own hearing on the Bhopal tragedy. Here the issue was worker safety in US plants that might pose risks similar to those at the Union Carbide plant in Bhopal. When that hearing made it clear that workers in the plant, as well as people living around the Union Carbide facility, had little or no knowledge about the toxicity of the chemicals used there, the subcommittee renewed its call for a stronger and expanded federal "right-to-know" protection for workers and communities around plants using dangerous chemicals. The Occupational Safety and Health Administration (OSHA) had issued rules in late 1983 setting standards for worker access to such information. The rules would not be fully effective until May 1986—too long to wait, in the eyes of some committee members in the wake of Bhopal. Moreover, OSHA regulations did not extend the "right to know" to surrounding communities.

Media coverage continued as the staggering numbers of the dead and injured in Bhopal climbed, but it was directed less to the human cost borne by the citizens of Bhopal than to preventing a similar tragedy on American soil. By February 1985, nearly a dozen bills had been introduced in Congress as a result of the Bhopal tragedy. One major package of bills was introduced by Florio shortly after the first set of hearings. Florio's package sought to amend five existing environmental laws, adding safeguards and including "right-to-know" provisions to give workers and communities information about the chemicals used at local plants as well as the amount and type of pollutants released to the environment.[74]

At the same time, Waxman was working with Representative Tim Wirth (D-CO) to introduce legislation that would change the way hazardous air pollutants were identified and regulated by the EPA. At the hearing, Waxman revealed that major US companies had identified some 204 chemicals they considered hazardous, almost all of which were emitted into the air, sometimes in very high quantities. "It is a sad commentary that no government agency has ever attempted to gather this information," Waxman said. "In fact, when we sought the help of EPA, we discovered that they do not even have an up-to-date list of where the chemical plants in this country are located."[75]

On March 26, 1985, during joint hearings held by House energy and commerce subcommittees, the chemical industry appeared to capitulate to public pressure for greater oversight and information. Union Carbide chairman Warren Anderson urged Congress to "improve regulatory control over hazardous air pollutants" under the Clean Air Act.[76] Another executive, Harold J. Corbett of the Monsanto Company, told the lawmakers: "There might be merit in federal legislation regulating emergency response" to accidents in chemical plants.[77] The Chemical Manufacturers' Association (CMA), while stopping short of endorsing any of the bills, did not publicly oppose the new bills, as they had done with similar measures in previous years. CMA spokesman Tom Gilroy equivocated in his testimony: "I think everyone in the industry recognizes that regulations are sometimes necessary if they're crafted correctly."[78]

A pivotal figure in moving legislation to the floor of the House was John D. Dingell (D-MI), chairman of the House Energy Committee. In the past, Dingell had provided a sympathetic ear to voices in the auto and other industries regarding additional regulations under the Clean Air Act, working to soften regulations and regulatory deadlines. In this situation, however, he understood that the political winds would force Congress to respond. "The demand for action is growing, particularly in light of the recent Bhopal disaster," Dingell observed during a speech before the CMA. Dingell noted: "Either the folks who know the industry can fashion, in a timely manner, a program to inform the public about operations, and a program to protect communities surrounding operations from emergency situations, or we in Congress, who know much less about the industry, will probably do it for you."[79]

A NEW LAW: THE EMERGENCY PLANNING AND COMMUNITY RIGHT-TO-KNOW ACT OF 1986

On May 22, 1985, three House Energy subcommittee chairs—Waxman, Florio, and Wirth—introduced H.R. 2576, the Toxic Release Control Act of 1985. In introducing the bill, Waxman wryly noted that the chemical industry was on a nationwide "honor system" that did not require reporting toxic releases. He observed that the EPA had neither an up-to-date list of chemical facilities nor a list of the chemicals emitted from those facilities. Waxman further noted that "most Americans would be staggered to know, in the face of today's concern with over-regulation, that the chemical industry is virtually free to release into the air unlimited quantities of whatever poisonous or cancer-causing chemicals that it sees fit."[80]

The imprint of Bhopal on the bill was clear. The bill required that chemical companies have vigorous maintenance schedules, minimal requirements for monitoring leaks, and a special permit requirement for "extremely hazardous substances" (such as methyl isocyanate or phosgene).[81] Section 301 of H.R. 2576 tasked the EPA with a careful review of operations at manufacturing and chemical facilities. In addition to establishing monitoring and leak control requirements, the agency would set equipment design, work practice, and operational requirements for chemical plants.

Other sections also invoked the specter of Bhopal. Section 203 of the bill delineated "community right-to-know" provisions, which required companies to identify the hazardous substances that could arise from their operations, along with the concentrations of those substances and potential adverse health effects. Section 601 dealt with emergency response planning and required manufacturers of dangerous chemicals to develop comprehensive evacuation and emergency response plans that included working with local and state officials. Additionally, local and state officials were to be notified immediately after a leak and informed of all measures that a company had taken to mitigate the leak and minimize risks to human health and the environment.

While the right-to-know and emergency planning provisions found support during hearings on the bill, the modification of Section 112 of the Clean Air Act did not. Among the most contentious language in the bill was the listing of eighty-five chemicals (including MIC) that would be regulated as hazardous air pollutants. Further, the language of the bill gave the EPA the ability to set technology-based standards rather than health-based standards. Congressional architects debated jurisdictional issues as well. Most notable were considerations of whether the right-to-know requirements were better placed in the bill reauthorizing the Superfund or in the Clean Air Act.

Over the next few months, members of Congress engaged in fierce debate over the parameters of the amendments to CERCLA and the CAA and the most appropriate place to put right-to-know provisions. Then, as described earlier, the toxic release from Union Carbide's plant in Institute, West Virginia, provided a second trigger for congressional action. This event sent 135 residents to area hospitals and confined thousands more to their homes, and it happened in spite of Union Carbide's statement that a leak like Bhopal's could not occur here, and despite the company's $6 million investment in additional safety systems at the plant. Perhaps most important, Union Carbide did not clearly and immediately warn the public when the Institute leak happened. With a leak of toxic gas on US soil, the stage for congressional action was set.

Over thirty bills would be introduced in 1985 dealing with Superfund reauthorization, many of them including disclosure and reporting requirements for firms using hazardous chemicals. On December 3, 1985—coincidentally the first anniversary of the Night of the Gas—a legislative compromise was reached in conference committee, setting the stage for the passage of the Superfund Amendments and Reauthorization Act in 1986. The bill that emerged from the committee contained tough new provisions for the chemical industry.

The bill that would eventually become law, H.R. 2005, emerged with right-to-know requirements intact, as well as emergency planning and response requirements. Also intact was congressional agreement that major industrial sources should be required to publicly report releases of chemicals into the air, land, and water. The bitterest divisions around the reauthorization of the main title in the Superfund act—for example, who would pay the multibillion-dollar price tag for cleaning hundreds of abandoned waste sites around the country—are not the topic of this chapter, but they do make for an interesting tale of policy formation. Noteworthy for our purposes here is the passage of Title III of the Superfund Amendments and Reauthorization Act (Pub. L. 99-499), known as the Emergency Planning and Community Right-to-Know Act of 1986 (EPCRA), on October 17, 1986.

EPCRA contained three major elements. Subtitle A developed emergency planning and notification requirements for facilities that used toxic substances. It required the EPA to identify and publish a list of extremely hazardous substances and threshold planning quantities for each substance. Facilities where such substances were present in such threshold quantities, as determined by the EPA, were required to identify an on-site emergency coordinator. The company's coordinator would also participate in the planning efforts of local emergency planning committees. Facilities would be required to notify these committees of accidental releases and inform them of any plant conditions that might pose a human or environmental health threat.

Subtitle A also placed requirements on state and local governments. The state would establish emergency response commissions to supervise and coordinate local emergency planning committees, which would develop and, when necessary, implement emergency response plans for hazardous substance emergencies within their jurisdiction. In short, Subtitle A was designed to ensure that communities were not caught unaware of potentially dangerous community exposures to toxic releases.

Subtitle B contained reporting requirements for covered facilities. Owners or operators of covered facilities were required to prepare a material safety data sheet (MSDS) for hazardous chemicals and provide that information to local and state emergency entities and appropriate fire departments. In a nod to the public's right to know about chemicals, MSDS information was to be made available to the public. Covered facilities would need to provide additional information about average inventories of toxic substances and the amount and storage of individual chemicals.

The most prominent reflection of the public's right to know about chemicals in local communities was the Toxics Release Inventory (TRI). This section of the new law required owners or operators of facilities to prepare, submit, and complete a toxic chemical release form on an annual basis, detailing the use, manufacture, presence, and disposal of listed toxic chemicals during that year. The EPA could revise the list of chemicals as appropriate to protect human health. This information, in turn, would be compiled by the EPA and placed in a national database that would be made available to the public, searchable by company name. This allowed any individual to know exactly how many chemicals were released into the environment by each company and to determine the long-term trends of toxic releases.[82] The TRI database is still a major source of information about toxins in communities today.

Advocates for addressing hazardous air pollution would eventually find legislative success by passing what was referred to as "the Bhopal amendment" in the 1990 Amendments to the Clean Air Act. Section 112(r) of the law required companies to disclose routine venting or discharge of chemicals that cause cancer, birth defects, and other chronic health problems.[83] It set out a basic statutory principle that companies are responsible for maintaining a safe facility, identifying hazards, and minimizing the consequences of accidental chemical releases.[84] Representative Bob Edgar (D-PA), one of Florio's allies, had fought for this change in 1985, noting, "The people we represent have a right to know if they are being exposed to chemicals that could potentially kill them—whether they die suddenly or over a decade."[85]

If Congress responded to the Bhopal tragedy, so did the EPA. In 1985, the agency started its Chemical Emergency Preparedness Program in response to public concern about a potentially serious chemical accident like Bhopal occurring in the United States. The program was voluntary, encouraging state and local officials to identify hazards and undertake emergency response planning. The following year, EPA established its Chemical Accident Prevention Program, designed to collect information about chemical accidents and also to encourage industries to improve safety at chemical facilities. These programs were reflected in the EPCRA law and the Clean Air Act amendments.

In sum, after the Bhopal tragedy provided a focusing event that prompted Congress to create new safeguards for American citizens, the second gas release in Institute, West Virginia, intensified public demand that Congress and the EPA act. With the passage of EPCRA, communities would have information on the hazardous chemicals present on-site in local facilities. The public would also have information about toxic substances being released into the environment by individual firms. Taken together, these elements, at minimum, provided a base of knowledge from which officials could respond to a chemical accident.

Congress was not successful, however, in curtailing how multinational corporations operated overseas. During the House Subcommittee on Asian and Pacific Affairs hearing, Reagan administration officials stuck by the long-standing policy of not making US corporations adhere to US environmental regulations while operating abroad, instead pointing to the ability of the host nation to establish its own environmental requirements.

CONCLUSION

Who is to blame for this "Hiroshima of the chemical industry," as it was later described by the Citizens Commission on Bhopal?[86] Was the explosion at the Union Carbide plant in Bhopal an unavoidable accident that could have happened to any pesticides manufacturer anywhere? That is to say, was the cascade of events in this high-technology company so interactively complex and tightly coupled that no one could have anticipated how these tragic consequences would unfold? The results of a host of studies—some of which have been reported in this chapter—suggest that the Bhopal disaster was anything but a "normal" or systems accident.

From the beginning of the investigations into the tragedy, Union Carbide steadfastly maintained that the processes and procedures it mandated for the Bhopal plant were sound and safe. Ronald Wishart, vice president for public affairs at Union Carbide, observed shortly after the accident that there was no health and safety double standard for Union Carbide plants operating outside of the United States.[87] This comment was later retracted in the face of clear evidence to the contrary.

Reconstruction of the disaster revealed that Union Carbide and UCIL made many bad decisions, most of them reflecting a lack of regard for on-the-ground safety and environmental protection and an atrophy of vigilance, even in the face of multiple previous leaks and fatal accidents. Investigations detailed that management practices at Union Carbide were consistently focused on cost-cutting measures that reduced or eliminated safety and environmental safeguards. Moreover, the attitudes of key decision-makers at UCIL border on villainous. For example, in a display of hubris at worst, or overconfidence at best, Union Carbide officials believed in their ability to control adverse consequences by under-engineering or shutting down at least four safety systems.

In his 1999 update to *Normal Accidents* (1984), Charles Perrow deems Bhopal a preventable disaster. He notes that the company paid little attention to any of the precursors, or warnings, in the system. After repeated leaks at the plant, the company continued to engage in risky behavior, even expanding its storage of MIC. In critiquing Union Carbide, Perrow describes what he calls the "Union Carbide Factor."[88] He warns that a potent combination of the following variables found in the Bhopal disaster would give rise to a future catastrophe: high quantities of toxic substances, ignorance of the toxic or explosive dangers, no warning systems, large numbers of unprotected people, a vulnerable environment, and a lack of immediate response to help victims.

It could be argued that the responsibility for the tragedy rests solely with UCIL, the Union Carbide subsidiary that operated the plant in Bhopal, since plant managers on the ground are certainly in the best position to oversee what is happening in production facilities. This is essentially what Union Carbide argued in legal battles and in the court of public opinion after the tragedy. Union Carbide insisted that its subsidiary was responsible for any substandard processes that were in place at the plant. UCIL, not surprisingly, countered that primary responsibility for the accident rested with Union Carbide. UCIL officials maintained during investigations into the cause of the accident that the design of the plant, the decisions made at the plant regarding storage of deadly chemicals, and the decisions to minimize costs by reducing the number and sophistication of employees all emanated from the US headquarters. Moreover, Union Carbide owned the controlling stake in its subsidiary and therefore controlled the culture at the plant, if not the daily operations.

And what of the Indian government? Did government actions in India (or actions of the US government, for that matter) serve to protect the citizens of Bhopal, or did they ultimately put citizens in greater peril? Government regulations are designed to watchdog the self-serving actions of business. Government has a responsibility to monitor companies in an effort to minimize risks to the populace it serves. Here it appears that the Indian government, eager to get and keep a large multinational corporation, was willing to change zoning requirements and ownership requirements and do whatever it could to maintain support for Union Carbide.

By any measure, Bhopal was a tragedy that keeps claiming victims—not once, but twice, and perhaps even into a third generation. It has altered the human and environmental mosaic of the community in ways that are almost too gigantic in scope to understand. It's as if a community of fifteen thousand people was wiped from the face of the earth on the Night of the Gas, and then ten or fifteen times that number of people have continued—again, not for a short time, but for years—to bear the costs of that night. It is telling that *Bhopali*, a 2011 documentary, talks about the next generation as the "children of Bhopal."

Although the media attention surrounding this unparalleled industrial disaster produced a new environmental law in the United States that alerts us to the chemicals in use in our communities, the question of how we should respond to the overseas actions of multinational corporations remains open. Should we expect plants located in other parts of the world to adhere to the same standards required of companies in the United States?

If the value of a human life is the same throughout the world, the answer seems obvious.

It is no surprise that people in Bhopal responded angrily to the $20 billion settlement offered by BP to people affected by its oil spill in the Gulf of Mexico in 2010 (see Chapter 3). Not only was that financial compensation offered shortly after the accident (the people of Bhopal waited twenty years for compensation), but the settlement was forty times larger than what was offered to those affected in Bhopal, even though the BP oil spill involved nothing close to the human cost of the Bhopal gas leak.[89] As the Indian government said on its website dedicated to the Bhopal disaster: "When some more time passes and people overcome their shock and grief and after acquiring perspective, when some[one] compiles an account of this tragedy, it will surely stand out in history as the most tragic, the most cruel and the most bizarre calamity perpetrated by man on man."[90]

DISCUSSION QUESTIONS

1. Do you think an event like the Bhopal gas leak could happen in the United States? Why or why not?
2. Should Warren Anderson have been extradited to India to face criminal charges? Why or why not?
3. What do you see as the most compelling evidence that what happened at Union Carbide's pesticides factory in Bhopal was not a "normal" accident?
4. Who bears responsibility for the environmental cleanup of a pesticides factory?

NOTES

1. Madhya Pradesh, Bhopal Gas Tragedy Relief and Rehabilitation Department, Bhopal, "Profile"; Bridget Hanna, Ward Morehouse, and Satinath Sarangi, eds., *The Bhopal Reader: Remembering Twenty Years of the World's Worst Industrial Disaster* (New York: Apex Press, 2005), 3.

2. For this claim put forth by Union Carbide, see Union Carbide Corporation, "History," www.unioncarbide.com/history/.

3. David Dembo, Ward Morehouse, and Lucinda Wykle, *Abuse of Power: Social Performance of Multinational Corporations: The Case of Union Carbide* (New York: New Horizons Press, 1990), 14.

4. Dembo, Morehouse, and Wykle, *Abuse of Power,* 16; Hanna, Morehouse, and Sarangi, *The Bhopal Reader,* 25.

5. "The Ghosts of Bhopal, 30 Years After Disaster," *Toronto Star,* November 21, 2014.

6. Union Carbide, Bhopal Information Center, "Bhopal Gas Tragedy Information," www.bhopal.com/.

7. Dembo, Morehouse, and Wykle, *Abuse of Power,* 76.

8. Hanna, Morehouse, and Sarangi, *The Bhopal Reader,* xxiv.

9. Pico Iyer, "India's Night of Death. Bhophal," *Time,* December 17, 1984.

10. Larry Everest, *Behind the Poison Cloud: Union Carbide's Bhopal Massacre* (Chicago: Banner Press, 1985), 12.

11. Raajkumar Keswani, "An Auschwitz in Bhopal," *Bhopal Post,* June 28, 2010, www.thebhopalpost.com/index.php/2010/06/an-auschwitz-in-bhopal/ (accessed December 11, 2010).

12. Hanna, Morehouse, and Sarangi, *The Bhopal Reader,* xxv.

13. Amy Waldman, "Bhopal Seethes, Pained and Poor 18 Years Later," *New York Times,* September 21, 2002.

14. Iyer, "India's Night of Death."

15. Government of Madhya Pradesh, Bhopal Gas Tragedy Relief and Rehabilitation Department, "Immediate Relief Provided by the State Government," http://demosl56.rvsolutions.in/ernet/website-050/immediate-relief.

16. Government of Madhya Pradesh, Bhopal Gas Tragedy Relief and Rehabilitation Department, "Facts and Figures," http://demosl56.rvsolutions.in/ernet/website-050/facts-figures.

17. Keswani, "An Auschwitz in Bhopal."

18. Sominin Sengupta, "Decades Later, Toxic Sludge Torments Bhopal," *New York Times,* July 7, 2008.

19. Union Carbide Corporation, "Remediation (Clean Up) of the Bhopal Plant Site," www.bhopal.com/Remediation-of-Bhopal-Plant-Site (accessed September 17, 2017).

20. Sara Goodman, "Poisoned Water Haunts Bhopal 25 Years After Chemical Accident," *Scientific American* (December 1, 2009).

21. Colin Toogood, "The Second Bhopal Disaster," *Crisis Response* 10, no. 2 (December 2014).

22. Ibid.

23. Tim Edwards, "Second Poisoning," The Bhopal Medical Appeal, 2014, http://bhopal.org/second-poisoning/ (accessed June 2, 2017). For a chronology of environmental studies of the site, see Union Carbide Corporation, "Environmental Studies of the Bhopal Plant Site," www.bhopal.com/Environmental-Studies-of-Bhopal-Plant-Site (accessed September 17, 2017).

24. Ibid.

25. Indra Sinha, "Bhopal: 25 Years of Poison," *The Guardian,* December 3, 2009.

26. I. Labunska, A. Stephenson, K. Brigden, R. Stringer, D. Santillo, and P. A. Johnston, *The Bhopal Legacy: Toxic Contaminants at the Former Union Carbide Factory Site, Bhopal, India: 15 Years After the Bhopal Accident,* Technical Note 04/99

(Exeter, UK: University of Exeter, Greenpeace Research Laboratories, November 1999), www.greenpeace.org/india/Global/india/report/1999/10/the-bhopal-legacy .pdf (accessed June 2, 2017).

27. Ibid., 4.

28. "Statement of The Dow Chemical Company Regarding the Bhopal Tragedy," www.dow.com/en-us/about-dow/issues-and-challenges/bhopal/dow-and-bhopal.

29. Jessica Haynes, "Dow: No Obligation in Bhopal Court Notice," *Midland Daily News,* December 21, 2015, www.ourmidland.com/news/article/Dow-No-obligation -in-Bhopal-court-notice-6904354.php (accessed June 2, 2017).

30. Shashikant Trivedi, "Bhopal Gas Tragedy—31 Years On: A Third Generation Might Be at Risk," *Business Standard,* December 2, 2015, www.business-standard .com/article/current-affairs/bhopal-gas-tragedy-31-years-on-a-third-generation -might-be-at-risk-115120200028_1.html (accessed September 19, 2017).

31. Shashikant Trivedi, "Decision on Toxic Waste on June 8 Center to Take over Bhopal Memorial Hospital," *Business Standard,* June 7, 2012, www.business -standard.com/article/economy-policy/decision-on-toxic-waste-on-june-8-center -to-take-over-bhopal-memorial-hospital-112060702015_1.html# (accessed June 2, 2017).

32. Trivedi, "Bhopal Gas Tragedy—31 Years On."

33. Ibid.

34. Ward Morehouse and M. Arun Subramaniam, *The Bhopal Tragedy: What Really Happened and What It Means for American Workers and Communities at Risk: Preliminary Report for the Citizens Commission on Bhopal* (New York: Council on International and Public Affairs, 1986), 3.

35. International Campaign for Justice in Bhopal, "Affidavit of Edward Munoz," Judicial Panel on Multidistrict Litigation, Docket 626, January 28, 1985 (source documents), www.bhopal.net/source_documents/munoz%20affidavit1985.pdf; see also www.bhopal.net/key-source-documents-in-the-unheard-case-against-ucc/.

36. Dembo, Morehouse, and Wykle, *Abuse of Power,* 87.

37. Hanna, Morehouse, and Sarangi, *The Bhopal Reader,* xxiv.

38. International Campaign for Justice in Bhopal, "Excerpt from May 1982 Safety Audit of the Bhopal Plant by US Engineers" (source documents), 6, http://bhopal .net/source_documents/1982%20safety%20audit.pdf.

39. Hanna, Morehouse, and Sarangi, *The Bhopal Reader,* 35.

40. Ibid., xxiv.

41. Everest, *Behind the Poison Cloud,* 57.

42. International Campaign for Justice in Bhopal, internal memorandum from Union Carbide Eastern, Inc., February 24, 1984, Exhibit 46, 2, http://bhopal.net /source_documents/UCE%20review%20of%20UCIL-exhibit%2046.pdf.

43. Hanna, Morehouse, and Sarangi, *The Bhopal Reader,* 30.

44. Dembo, Morehouse, and Wykle, *Abuse of Power,* 87.

45. Hanna, Morehouse, and Sarangi, *The Bhopal Reader,* 32.

46. Everest, *Behind the Poison Cloud,* 21.

47. Ibid., 22.

48. International Program on Chemical Safety (IPCS), "Methyl Isocyanate," validated November 27, 2003, www.inchem.org/documents/icsc/icsc/eics0004.htm (accessed December 10, 2010).

49. Ravi Kiran and Shamanth Jilla, "7 Ways the Government Played a Role in the #BhopalGasdisaster," Kractivism, March 12, 2014, www.kractivist.org/7-ways-the-government-played-a-role-in-the-bhopalgasdisaster/ (accessed June 1, 2017).

50. Morehouse and Subramaniam, *The Bhopal Tragedy*, 3.

51. Hanna, Morehouse, and Sarangi, *The Bhopal Reader*, 14.

52. Keswani, "An Auschwitz in Bhopal."

53. Quoted in Hanna, Morehouse, and Sarangi, *The Bhopal Reader*, 17; see also Keswani, "An Auschwitz in Bhopal."

54. Keswani, "An Auschwitz in Bhopal."

55. Shahnawaz Khan, letter to Union Carbide India Limited, March 4, 1983, reprinted in Hanna, Morehouse, and Sarangi, *The Bhopal Reader*, 19.

56. Sheila Jasanoff, "Bhopal's Trials of Knowledge and Ignorance," *Isis* 98 (2007): 344–350.

57. S. Muralidhar, *Unsettling Truths, Untold Tales: The Bhopal Gas Disaster Victims "Twenty Years" of Courtroom Struggles for Justice* (Geneva, Switzerland: International Environmental Law Research Centre, 2004), 10, www.ielrc.org/content/w0405.pdf (accessed on December 2, 2010).

58. "Indian Bare Acts: The Bhopal Gas Leak Disaster (Processing of Claims) Act, 1985" (March 29, 1985), Help Line Law, www.helplinelaw.com/docs/the-bhopal-gas-leak-disaster-processing-of-claims-act-1985.

59. Ibid.

60. Muralidhar, *Unsettling Truths, Untold Tales*, 12.

61. Madhur Singh, "Bhopal Victims Still Seeking Redress," *Time*, June 28, 2008.

62. Peter Stoler, Dean Brelis, and Pico Iyer, "India: Clouds of Uncertainty," *Time*, December 24, 1984.

63. Ibid.

64. Amnesty International, "First Convictions for 1984 Union Carbide Disaster in Bhopal Too Little, Too Late," June 7, 2010, www.amnesty.org/en/latest/news/2010/06/first-convictions-1984-union-carbide-disaster-bhopal-too-little-too-late/.

65. Lydia Polgreen and Hari Kumar, "8 Former Executives Guilty in '84 Bhopal Chemical Leak," *New York Times*, June 7, 2010.

66. Rubina Khan Shapoo and A. Vaidyanathan, "Supreme Court Reopens Bhopal Gas Tragedy Case," *NDTV*, August 31, 2010, www.ndtv.com/article/india/supreme-court-reopens-bhopal-gas-tragedy-case-48514?cp (accessed August 14, 2010).

67. Ibid.

68. NDTV correspondent, "Bhopal Gas Tragedy: Supreme Court Rejects CBI Plea to Re-open Case," *NDTV*, May 11, 2011, www.ndtv.com/india-news/bhopal-gas-tragedy-supreme-court-rejects-cbi-plea-to-re-open-case-455478 (accessed June 2, 2017).

69. Press Trust of India, "Bhopal Gas Tragedy Is a Closed Case Now: US," *NDTV,* August 20, 2010. www.ndtv.com/article/world/bhopal-gas-tragedy-is-a-closed-case -now-us-45797 (accessed October 14, 2010).

70. Joseph A. Davis and Nancy Green, "Bhopal Tragedy Prompts Scrutiny by Congress," *CQ Weekly Online,* December 22, 1984, 3147–3148, http://library.cqpress .com/cqweekly/WR098404004 (accessed October 20, 2010).

71. These criteria pollutants are carbon monoxide, lead, sulfur dioxide, ozone, particulates, and nitrogen oxides.

72. Davis and Green, "Bhopal Tragedy Prompts Scrutiny by Congress," 3147.

73. Ibid., 3148.

74. Joseph A. Davis, "Some in Chemical Industry Endorse Regulation: Bills on Toxic Pollutants Picking Up Support," *CQ Weekly Online,* March 30, 1985, 602, http://library.cqpress.com/cqweekly/WR099404457 (accessed September 19, 2017).

75. Quoted in ibid., 602.

76. Ibid.

77. Ibid.

78. Ibid.

79. Ibid.

80. Joseph A. Davis, "Major Bill on Toxic Air Pollution Introduced," *CQ Weekly Online,* May 25, 1985, 1010, http://library.cqpress.com/cqweekly/WR099404758 (accessed October 20, 2010).

81. James A. Merchant, "Preparing for Disaster" (editorial), *American Journal of Public Health* 76, no. 3 (March 1986): 233–235.

82. Michael Kraft, Mark Stephan, and Troy Abel, *Coming Clean: Information Disclosure and Environmental Performance* (Cambridge, MA: MIT Press, 2011).

83. Joseph A. Davis, "Compromise Reached Between Committees: House Debates $10 Billion Bill for Hazardous-Waste Cleanup," *CQ Weekly Online,* December 7, 1985, 2552–2553, http://library.cqpress.com/cqweekly/WR099405861 (accessed October 20, 2010).

84. EPA, "Chemical Accident Prevention and the Clean Air Act Amendments of 1990," Fact Sheet 550-F-96-004, May 1996.

85. Quoted in ibid., 2553.

86. Morehouse and Subramaniam, *The Bhopal Tragedy,* vii.

87. Dembo, Morehouse, and Wykle, *Abuse of Power,* 78.

88. Charles Perrow, *Normal Accidents: Living with High-Risk Technologies* (New York: Basic Books, 1984; updated edition, Princeton, NJ: Princeton University Press, 1999), 359.

89. It should be noted that this settlement has been called a "down payment" on the spill's aftermath and does not include any fines or forfeitures that BP may be assessed, nor does it include any costs of litigation or damages that may come from litigation, as will be described in Chapter 3.

90. Government of Madhya Pradesh, Bhopal Gas Tragedy Relief and Rehabilitation Department, http://demosl56.rvsolutions.in/ernet/website-050/.

Deep Trouble

THE BP OIL SPILL

April 20, 2010, began as an unremarkable day for people around the Gulf of Mexico coast. About forty-nine miles off the Louisiana coast, a gigantic drilling rig, the *Deepwater Horizon,* was finishing its work on the Macondo well, an exploratory well that lay under five thousand feet of Gulf water and extended another thirteen thousand feet under the seafloor. The *Deepwater Horizon* was a modern-day technological marvel: looming over three hundred feet high, it had a deck almost as big as two football fields. Home for a crew of 126, it operated twenty-four hours a day. Earlier that day, engineers from Halliburton, a company working for BP, the oil company running the operation, had just finished cementing the well's final casing.[1] This plug was intended to keep the Macondo well stable until BP returned with an oil production platform from which it would harvest the crude oil. Its work nearly done, the *Deepwater Horizon* was getting ready to be moved to a new location.

Owned by Transocean, Inc., the *Deepwater Horizon* had been leased to the London-based oil giant BP since 2007. The Switzerland-based Transocean rented rigs, equipment, and personnel to major oil companies for deep-sea drilling operations, and BP was one of its best customers. Unlike rigs of the past, which used anchor mooring systems, the *Deepwater Horizon* represented state-of-the-art dynamic positioning technology to keep it in a precise location. This sophisticated system relied on computer-controlled thrusters that kept the rig in position above the well and the well connected to the rig by just a riser—a pipe that allows the circulation of drilling mud from the rig to the well. This was an enormous semi-submersible rig, a kind of ship that, instead of being anchored to the seafloor, floated on the surface of the water. This kind of technology was a necessity for drilling in increasingly deep water.

The *Deepwater Horizon* was drilling the Macondo well in a part of the Gulf known as the Mississippi Canyon Block 252. Just a few months before, the *Deepwater Horizon* had completed the deepest offshore exploratory well in history, reaching a depth of 35,055 feet, in a BP discovery called Tiber Prospect.[2] As the largest producer of oil and gas in the Gulf, with a net production of over four hundred thousand barrels a day, BP was proud of its accomplishments.[3] The *Deepwater Horizon* was going to help BP further explore drilling opportunities in the Gulf, and the future looked promising.

Without a doubt, BP and Transocean were on a roll: BP would find the oil, and Transocean had the drilling rig to get to it on the seafloor. As Transocean boldly asserted in its corporate motto, "We're never out of our depth."[4] This statement would come back to haunt Transocean, the world's largest offshore drilling contractor, and BP, one of the world's preeminent international oil companies, on April 20, 2010. Late that evening, a blowout at the base of the well sent oil and gas surging up to the *Deepwater Horizon*, releasing gas that prompted a series of explosions that created a fireball of the *Deepwater Horizon* and killed eleven crew members. Thus began the worst environmental disaster in US history.

This chapter tells the story of BP and the cataclysmic failures that caused the world's largest oil spill. To explore what caused the deadly explosion, the chapter provides an overview of the emergence of BP against the backdrop of a country hungry for oil and an oil company hungry for profits amid competition from other major international oil companies. We then explore how both government and industry not only failed to anticipate and prevent this catastrophe but also failed to be prepared to respond to it. The common characteristics of industrial disasters—a history of disregard for safety and environmental standards, a myopic focus on the bottom line, lack of sufficient planning and preparation, and an ineffective regulatory presence (see Box 1.1)—permeate the BP oil spill story. As the National Commission on the BP *Deepwater Horizon* Oil Spill and Offshore Drilling noted in its report, this story has "recurring themes of missed warning signals, failure to share information, and a general lack of appreciation for the risks involved."[5]

Photo 3.1: Fireboat response crews battle the blazing remnants of *Deepwater Horizon*. Multiple Coast Guard helicopters, planes, and cutters responded to rescue the *Deepwater Horizon*'s 126-person crew. *US Coast Guard.*

This story includes the enormous costs of the disaster, in both financial and environmental terms. It also describes how the disaster upset the balance of regulatory approaches to offshore drilling, suggesting that the BP oil spill is an example of "punctuated equilibrium" that prompted policy change. The chapter concludes by looking at America's addiction to oil, US energy policy, and how the rest of us can be part of this continuing story.

BLOWOUT: THE *DEEPWATER HORIZON* IN FLAMES

At 9:52 p.m. on April 20, 2010, the *Deepwater Horizon* issued a mayday call to the US Coast Guard. Captain Curt Kuchta was initially not happy with Andrea Fleytas, the dynamic positioning officer who had made the call without authorization.[6] The captain would soon understand, however, that Fleytas had made the right call. He now commanded a modern-day *Titanic*. Just two minutes earlier, Doug Brown, the rig's chief mechanic, had heard a deafening hiss followed by an explosion that knocked him into the engine console. The hiss was the sound of thousands of cubic feet of natural gas coming from three miles beneath the deck of the rig. Gas coming from the well was being sucked into the air intakes of the massive deck engines that powered the rig. In moments, the combination of fuel and hydrocarbons overwhelmed engine number 3, which exploded with mighty force. *Deepwater Horizon* experienced two explosions in less than a minute.[7]

The exploding deck engine had set off a runaway reaction, igniting the giant stream of oil and gas coming up through the rig floor and incinerating the derrick tower some 240 feet above the deck. The explosions blew out the three-inch-thick steel doors and opened holes in the floor of the deck. Mike Williams, an electronics engineer, was hit by the first fire-safe door that protected the electronics shop. He crawled toward the second fire door, only to have that door explode off its hinges and hurtle into him. "And I remember thinking to myself, *This is it. I'm going to die right here,*" Williams would later recount to Scott Pelley in a CBS interview.[8] He and Brown made it to a deck that now resembled a combat zone.

In a tragic example of Murphy's Law, virtually everything on the *Deepwater Horizon* that could go wrong did. Few safety features functioned as planned. The blowout preventer failed, allowing the eruption of oil and gas and drilling mud to burst from the Macondo well. On the rig itself, the dynamic positioning system was down. The thrusters were down. Phone lines were down. The alarms were not set. The engines were disabled. The only hope the crew had for saving the *Deepwater Horizon* was to activate the emergency disconnect system to get the rig away from the blown-out well. This, too, proved fruitless. The system indicated that the emergency disconnect had happened, but the badly disabled rig could not float free of the well, and the raging inferno continued. At that point, the call came to abandon ship.

Workers abandoned the rig any way they could. Some left on lifeboats or emergency rafts, while others, like Williams, having missed the two departing lifeboats, grabbed life jackets and jumped into the water ten stories below. A supply ship in service to the *Deepwater Horizon,* the *Damon B. Bankston,* offered refuge for the survivors found in the dark water. Though the US Coast Guard searched for the next three days, eleven members of the crew would never be found. Seventeen more were seriously injured. Without the *Bankston*'s rescue craft nearby to save crew members in the water, casualties would probably have been much higher.[9]

The seemingly invincible *Deepwater Horizon* was now an inferno that burned for thirty-six more hours, drifting and listing, before succumbing on April 22 (coincidentally Earth Day) to its own watery grave nearly a mile below the surface and 1,300 feet from the blown-out well. But this was far from the end of the story. As the rig sank, the riser, which had never successfully been disconnected during the emergency procedures, kinked and fell to the bottom. As the riser fell, oil and gas gushed out of several breaks in the pipe, putting more stress on the failed blowout preventer. The world would watch with horror as oil erupted out of the Macondo well unabated for eighty-seven days. When the well was finally capped, at least 3.19 million barrels of oil—roughly 134 million gallons—had spilled into the Gulf.[10] Dwarfing the *Exxon Valdez* spill in 1989 of just over 260,000 barrels of oil (10 million gallons) into Alaska's Prince William Sound, the *Deepwater Horizon* spill was instantly the biggest US oil spill in history.

Prior to the accident, the Macondo well proved challenging for workers on the *Deepwater Horizon*. In October 2009, BP had started drilling the Macondo well with another rig, the *Marianas*. However, the older *Marianas* used a mooring system and proved unable to move easily out of the path of Hurricane Ida later that year. The damaged *Marianas* was replaced by the *Deepwater Horizon* in January 2010. But replacing the rig did not stop the problems. The crew on the *Deepwater Horizon* experienced four serious well-control events prior to the blowout in April 2010.[11]

A constant concern was the tendency of the well to "kick," or release pressure, which took its toll on the crew. In mid-March, the drill pipe became stuck, a dangerous condition in deepwater drilling; the crew was forced to drill a sidetrack hole. During testimony before the Joint Investigation Committee, BP drilling engineer Mark Hafle admitted to these major problems, and Mike Williams told committee members that the crew had dubbed the Macondo the "well from hell."[12] Drilling Macondo had proved complicated, requiring that engineers modify their plans as they learned more about the geologic formations thousands of feet below. Without a doubt, Macondo lived up to its reputation whether measured by loss of life, economic losses, or environmental damage.

Initially, BP's CEO, Tony Hayward, was confident that handling the sinking of the *Deepwater Horizon*, the failure of the Macondo well, and the subsequent leaking of oil into the Gulf were all well within BP's capabilities. When the *Deepwater Horizon* sank, it left a one-by-five-mile oil slick in its wake.[13] This was a concern, most observers thought, but certainly not a catastrophe. BP believed the leak to be small. Two days later, the Coast Guard, which was supervising the cleanup operations, estimated that two leaks from the blown-out well were spewing out as much as one thousand barrels, or forty-two thousand gallons, per day.[14] On April 28, the National Oceanic and Atmospheric Administration (NOAA) increased the official estimate to five thousand barrels of oil per day as BP noted a third leak from the riser. That figure greatly underestimated the true flow of oil from the well, which was later determined by scientists to have been fifty thousand to seventy thousand barrels per day—more than ten times that amount.[15]

This underestimate of the flow rate did little to help the BP engineers arrive at a workable solution. The flow rate helps determine the optimal design for well interventions. Also, the flow rate helps responders know how much dispersant should be used to address the oil slick. Planning for the containment of the oil at the surface is also dependent on knowing the amount of oil being released. Finally, the flow rate helps determine the rate at which the reservoir of oil beneath the seabed is being depleted. This helps determine the final shut-in pressure when the capping stack, an interim measure to plug the leak, is placed on top of the well.[16] Scientists study the pressure readings from the cap to determine the integrity of the well casing below the seafloor.[17]

By April 29, the spill was 120 miles long and threatened the Louisiana coast, prompting Homeland Security secretary Janet Napolitano to declare the spill an event of national significance. Meanwhile, environmental groups warned that the spill would have devastating effects on marine life along the coastal areas of Louisiana, Mississippi, Alabama, and Florida. President Barack Obama took his first trip to the area on May 2. Meeting with state and local officials, he observed that the leak might not be stopped for several more days—a wildly optimistic prediction, as it turned out.[18] The following day, Hayward promised that BP would pay for the cleanup and compensate people for legitimate losses due to the oil spill. It was an appropriate gesture, given the dire situation, but it was unlikely that he anticipated the full extent of the cost. Just one day later, oil reached the barrier islands off Chandeleur and Breton Sounds along the Louisiana shore.

The images of oil reaching shore were striking, but the live feed of oil gushing from the bottom of the seafloor had Americans riveted to their television sets. The live picture came from BP, which needed the cameras to operate in such deep water. The Macondo well was now a prime-time star. Americans watched in fascination and horror as remotely operated vehicles (ROVs) provided a bottom-of-the-sea view of the attempts to stop the gush of oil from the blown-out well. Multiple screens in a room called the HIVE (Highly Immersive Visualization Environment) at the oil giant's headquarters in Houston clearly showed that the oil spewed unabated from the seafloor, in quantities that were unimaginable to most viewers. At a mile below the surface, it was too risky for humans to attempt a repair. Instead, BP had to rely on unmanned vehicles and technical fixes. BP's attempts at stopping the flow of oil failed spectacularly on television screens across the country.

In its first attempt to control the well, BP used ROV intervention to engage the shear ram, which was designed to pinch off the pipe of an out-of-control well. This maneuver failed because the damage to the blow-out preventer and the presence of two pieces of drill pipe inside the stack made sealing it impossible.[19] Another early attempt by BP was to place a containment dome—a solid steel box four stories high and weighing nearly one hundred metric tons—over the top of the gushing well. This attempt failed almost immediately, however, when ice-like hydrate crystals formed inside the dome, clogging the riser pipe and even lifting this monstrosity off the seafloor. Instead of containing the flow of oil, the structure floated up toward the rescue boats on the surface. Had attempts to steer it away on May 7, 2010, failed, a collision between a ship and the methane-containing con-traption might have caused an explosion similar to the one that ultimately destroyed the *Deepwater Horizon*.[20]

A second, much smaller containment dome—dubbed a "top hat"—ar-rived at the scene on May 12, 2010. BP engineers would be able to pump methane into this smaller dome to prevent hydrate crystals from forming. That attempt was never made, however, perhaps because BP was coming to understand that the flow from the well was much greater than previously estimated. Such a small structure might have worked if the estimate of five thousand barrels per day had been accurate, but there was no way the "top hat" could contain fifty thousand barrels per day.

The third plan was to insert a tube into the wrecked riser to divert part of the flow to a vessel called the *Enterprise*. Though this strategy worked, it captured only two thousand barrels a day, just a fraction of the spewing oil. By this time, the oil plume spreading over the Gulf had prompted NOAA to close forty-six thousand square miles of federal waters to fishing.[21]

On May 26, BP technicians began firing heavy drilling mud into the well, using large thirty-thousand-horsepower pumps in a procedure known as "top kill." By forcing large amounts of heavy drilling mud into the well, the pressure would be reduced and the blown-out well could then be ce-mented. That effort had to be suspended three days later after BP deter-mined that too much of the mud was escaping out of the breach instead of going down the well. During this time, BP fired "junk shots" at the gushing well—a mix of golf balls, rubber balls, rope, and other debris—in an effort to slow down the flow of oil. The junk shots failed to slow the flow—in front of cameras for all to see—because the flowing oil was under pressure too great to be halted by any debris or mud. Nearly six more excruciatingly slow weeks would pass before BP finally succeeded in stopping the flow of crude oil at the Macondo well.

The final effort involved a complicated series of maneuvers. After the damaged riser pipe was sawed off using a remotely controlled diamond saw and shearing blade, the way was cleared for a new dome to be fitted over the top component of the blowout preventer. The final containment dome was set on July 12, and on July 15 oil stopped flowing into the Gulf for the first time since April 20. On September 3, the damaged blowout preventer was taken to the surface and given to the US Department of Justice as evidence in its investigation of the accident. Then a new blowout preventer was installed. On September 19, the "well from hell" was finally pronounced "dead."

AN ACCIDENT WAITING TO HAPPEN: IRRESPONSIBLE COMPANY CULTURE

The Macondo well blowout is hard to consider a "normal" accident. Though drilling a well almost a mile beneath the water's surface requires the use of an interactively complex and tightly coupled technological system (the factors identified by Charles Perrow as implicated in normal accidents), the BP oil spill is more than a story of technical failures or human error.

A Myopic Focus on the Bottom Line

Like Union Carbide, BP had a corporate DNA focused on the bottom line, which took precedence over safety or environmental considerations. Investigations, by both the federal government and BP itself, revealed that BP took shortcuts, failed to replace faulty equipment, and discouraged employees from pointing out potential safety issues. Even though it was BP policy that anyone with safety concerns could stop a project, the high number of accidents and fines suggests that employees understood that efficiency was the overriding pursuit. Jordan Barab, OSHA's deputy assistant secretary of labor for occupational safety and health, complained during a *ProPublica/ Frontline* interview: "They just weren't getting it. . . . BP's cost-cutting measures had really cut into their plant maintenance, into their training, into their investment in new and safer equipment." He continued, "When you start finding the same problems over and over again, I think you are pretty safe in saying they've got a systematic problem."[22]

The "well from hell" might have been tamed, given enough care and time. But taking time to proceed carefully was not a priority on the *Deepwater Horizon*. Daily operating costs were nearly $1 million, including the $500,000 BP paid Transocean each day it used the rig. By the time the well blew, BP was nearly $43 million over budget and forty-three days behind schedule.[23] Escalating costs had led BP to skip time- and money-consuming tests and safety procedures in what would be the final hours on the *Deepwater Horizon*, just as it had done in the months and years prior to the blowout. One such procedure was to establish a cement bond log to measure the effectiveness of cement seals on the Macondo's lateral walls, a test that should have been performed by the oil-field services contractor, Schlumberger Ltd. Schlumberger staff were on the rig and ready to run the test, but they were sent home on the morning of April 20, 2010, because BP wanted to save time and the $128,000 fee.[24] In perhaps the height of irony, Schlumberger staff left to make room for some BP executives coming to visit *Deepwater Horizon* in part to celebrate seven years of no lost-time accidents on the rig.[25] Had Schlumberger's crew run the test, they would have discovered that the newly cemented well was not properly sealed.

When the drilling was finished on the Macondo well, the next step was to finish the raw wellbore so as to prevent the highly pressurized oil and gas some thirteen thousand feet beneath the seafloor from "kicking," or releasing oil. Instead of employing a tieback liner, which would have provided extra protection to prevent a blowout, BP chose the cheaper single-string, or "long-string," design. The problem with this design is that it leaves a single pathway for a blowout all the way to the top of the well. Other methods provide multiple barriers to gas flowing at the well bottom. Other oil company executives testified that they preferred to use the tieback liner, which makes it possible, as a senior offshore design engineer from Shell explained, to "design your well so that you don't have to rely on blowout preventers for well control."[26]

When pressed during testimony before government investigators about the decision to use the long-string design, John Guide, the well team leader, denied that BP had chosen this potentially risky type of well casing over more traditional equipment because it would save the company three days and between $7 million and $10 million.[27] Tony Hayward, when asked about the selection of the long-string alternative during a congressional hearing, responded that it was not an unusual well design for deepwater drilling—a claim that was challenged by Halliburton staff, who stated that the cheaper, faster design was used in only one out of ten wells.

Choosing the cheapest alternative, one might argue, represents a sound business decision. In this case, however, it was a fatal one. BP, in choosing the long-string design, should have, at minimum, secured it by using centralizers. Herein lies the third choice that doomed the rig. Centralizers are rings that fit around the pipe, with braces that hold against the well wall and hold the casing string in the middle. These are critical components in ensuring a good cement job and keeping the cement even. Halliburton's casing and cement engineer for the Macondo well, Jesse Gagliano, was already concerned about the long-string option. He ran computer simulations and told BP that it would need more than six centralizers (the number then available to the *Deepwater Horizon*) to sustain the long-string casing. He indicated that with just six, the risk of gas flowing out of the well was "severe."[28] Instead of deferring to Gagliano's analysis, BP engineers thought that they would "probably be fine" without the additional centralizers.[29] As a consequence, the cementing process did not secure the casing, which shifted to the side and provided an opening for gas to leak.

Yet another choice played a role in the sinking of the *Deepwater Horizon*. On the morning of April 20, BP ran pressure tests to determine the well's integrity—a critical operation to ensure that a well is safe for abandonment. The negative pressure test, which measures upward pressure from a well, indicates its stability. Here again, three tests indicated that there was pressure in the well—a telltale sign that Macondo was seriously compromised. Instead of heeding the warning, BP officials chose to run a fourth test through the kill line, a separate pipe running from the drilling rig to the bottom of the blowout preventer. What they did not know was that a blockage in the kill line accounted for a successful negative pressure test. Inexplicably, BP and Transocean crew members confirmed the well's integrity based on the fourth test and chose to see the results from the first three tests as anomalies.

There were other contributing factors. BP chose a lightweight, nitrified cement not typically used as completion cement.[30] Also, because they opted to take the drilling mud out of the system too soon, the mud barrier was not there to stem the gas kick that destroyed the *Deepwater Horizon*. The size of the pipe (seven inches in diameter) and the size of the last section of the hole (eight and a half inches in diameter) left very little room for a cement sheath to stabilize the well. Another factor that may have contributed to the disaster was the failure to circulate the mud in a process known as "bottoms up," which helps to ensure that gas is out of the well and that the mud is clean of debris.

BP also opted to install one blind shear ram rather than two. Like Superman, the blind shear ram is meant to save the day in an emergency. When the Macondo's fury took hold of the *Deepwater Horizon,* the blind shear ram's two strong blades should have sliced through the drill pipe and sealed the well. But that fail-safe device didn't work. Investigators would later discover that the force of the blowout caused a section of pipe to bend and get stuck in the blowout preventer, disabling the blind shear rams.[31]

Since BP had opted to install only one blind shear ram, there was no backup. In its statement during the investigation, BP argued that space limitations on the rig had prohibited it from adding a second blind shear ram to the blowout preventer. However, experts told the *New York Times* that a second blind shear ram could have replaced some other component, and most of Transocean's other rigs had two.[32]

After the blowout, BP argued that blame should be placed on Transocean, the rig's owner. In an eerie parallel to actions taken by Union Carbide at the Bhopal plant, *Deepwater Horizon*'s alarm system was set to "inhibit" and thus did not sound on that fateful night. Although an operating alarm system would not have prevented the blowout, it would have alerted crew to the danger on the rig. Transocean claimed that this was not a safety oversight, but that workers were allowed to set the alarm to prevent it "from sounding unnecessarily when one of the hundreds of local alarms activates for what could be a minor issue or a non-emergency."[33] The *Deepwater Horizon* had a history of mechanical errors, which were documented by BP in a confidential 2009 audit. According to a review by the *New York Times,* BP officials discovered that Transocean had left 390 repairs undone—including many high-priority repairs—and that previously reported errors had been ignored by Transocean.[34] In a statement, BP made it clear that Transocean shared responsibility for the disaster: "As we have previously said, the *Deepwater Horizon* tragedy had multiple potential causes, including equipment failure."[35] As described later in the chapter, the courts were not persuaded.

A History of Complacency: Disaster Foretold in BP's Can-Do, Must-Do Culture

The accident on the *Deepwater Horizon* was not the company's first serious accident. By 2010, BP had a reputation among EPA and OSHA regulators in the United States as a renegade company, owing to a series of accidents and investigations. This "bad boy" reputation began in 1995, when the new CEO, Lord John Browne, was determined to build BP into the darling of Wall Street and make it a giant among oil companies. Believing that the best way to do this was by absorbing existing smaller oil companies, he took the company on a merger spree. In 1998, BP engineered a $61 billion buyout of Amoco Corporation—one of the largest industrial mergers in US history. Over the next two years, BP acquired ARCO and four other companies. US and British stock analysts cheered the new direction for BP and rewarded it with positive reviews. The stock price soared.

BP was now visible to the American public. It became the largest retailer of gasoline in the United States and second only to Exxon in market value.[36] Browne saw BP as the "anti-Exxon" oil company and highlighted that BP was not just an acronym for its long-standing name of British Petroleum but stood for "Beyond Petroleum," implying that the company would embrace a green economy. When BP chose not to challenge the Kyoto Protocol or the science behind climate change, it gained supporters in the environmental community.[37] This support eroded once BP's dismal safety and environmental record became known.

In his zeal to grow the company through acquisitions, Browne sought major efficiencies from these mergers, slashing jobs and wringing expenses out of existing equipment and maintenance operations. With an eye toward maximizing shareholder value by cost-cutting, Browne created a culture of corporate austerity and an ethos of corporate efficiency—with grave consequences for the people who worked for BP and for the environment.

In 2005, tragedy struck BP's oil refinery in Texas City, Texas, the third-largest oil refinery in the United States. On March 23, workers were restarting the isomerization unit, which helps improve the octane level in gasoline. Abnormal pressure was created in the production tower, triggering three relief valves. The relief valves opened to permit the gas to escape into a container called the blowdown drum. When the capacity of the drum was quickly overwhelmed, highly explosive liquid hydrocarbon spewed like a geyser into the air from a tower 120 feet overhead.[38] The liquid evaporated, creating a cloud of highly flammable gas that ignited, most likely after contact with an idling truck. The fireball explosion killed fifteen people and injured nearly two hundred more. Some 43,000 residents were ordered to remain indoors, and the forceful explosion damaged houses three-quarters of a mile from the refinery.[39] In addition to the human toll, the explosion heavily damaged the isomerization unit and the satellite control room, destroying the warehouse, vehicles, and mobile trailers as well as more than fifty storage tanks, resulting in financial losses exceeding $1.5 billion.[40]

The tragedy at the Texas City refinery prompted the largest investigation in the history of the US Chemical Safety and Hazard Investigation Board (CSB) and generated the largest OSHA fines ever levied. BP's own investigation pointed to operator error as the cause of the accident. The CSB disagreed, concluding that "safety system deficiencies created a workplace ripe for human error to occur."[41] In its report, the board charged that BP focused on the individual worker injury rate, but paid little attention to the overall safety culture, leaving the refinery's infrastructure and process equipment in disrepair. Inadequate operator training, a reduced training budget, and staff cuts were all part of BP's cost-cutting effort that left the refinery "vulnerable to a catastrophe."[42]

Months before the accident, an internal BP safety report seemed to agree with the conclusions of the investigators. The report found that cost-cutting had gotten to a critical stage and that employees had "an exceptional degree of fear of catastrophic incidents."[43] In response, BP spokesman Neil Chapman promised that the Texas refinery loss was "an event that has changed the company, and the commitment to safety has been reinvigorated."[44]

OSHA also was the subject of the CSB's scathing criticism. The board observed that OSHA conducted inspections of previous accidents involving fatalities at the refinery but never sounded the alarm about the likelihood of a catastrophic accident, nor did the agency conduct a comprehensive inspection of the process units. Only after the refinery explosion did OSHA move forcefully against BP, uncovering 301 egregious, willful violations at the Texas City refinery.[45]

After causing one of the worst industrial disasters in recent US history, BP should have gone on high alert to examine its operations and look for system deficiencies. However, just a year after the Texas City catastrophe, BP's failure to replace a corroded section of its pipeline in Alaska caused a spill of 212,000 gallons of crude oil—the largest spill ever on the Northern Slope. Whistleblowers had alerted the EPA that the pipeline was not being properly maintained. They said that the company policy was to minimize maintenance to squeeze out the maximum possible production, a policy referred to as "run to failure." During questioning by the government, BP admitted that it had gone eight years without "pigging" the pipeline—the standard maintenance process of running a mechanical robot through the pipe to clean it and measure corrosion—in part to save money.[46] When a second leak was discovered two days later, and with growing concerns about pipeline integrity, BP cut off the entire flow of oil from the North Slope overnight, sinking its stock and sending oil prices skyward.

As bad as these two incidents were, they were not the only accidents at BP facilities. They were just the most public. BP was subsequently charged with four federal crimes and "debarred" from government contracts at its Prudhoe Bay and Texas City operations. Debarment is a sanction available to federal officials after serious violations of federal law—in this case, the Clean Water Act. BP's debarment meant that these two operations could not do business with the federal government. Jeanne Pascal, a senior debarment attorney at the EPA, noted that BP was in a "league of its own" in its failure to come into compliance, and she had tried to pursue a debarment across all BP operations.[47] In 2009, she warned BP executives and EPA officials that the company's approach to safety and environmental protection made another disaster likely.

This warning should have come as no surprise to federal officials or BP executives. Between 2000 and 2009, BP had the dubious distinction of leading the rest of the industry in spills and serious safety violations. An analysis by ProPublica found that in Alaska the company produced about twice the amount of oil as ConocoPhillips, the other major oil company operating there, but recorded nearly four times as many large spills of oil, chemicals, or waste. In the Gulf of Mexico, BP had more spills than Shell, even though Shell produced more oil.[48]

BP's record is most abysmal regarding worker safety. OSHA recorded 518 safety violations between 1990 and 2009 for BP, more than twice the number recorded for Chevron (240) and almost five times as many as Exxon accumulated (108) over the same period. Later that year, OSHA slapped BP with another 745 violations at two refineries, including the one at Texas City. BP's poor safety record cost it more than $108 million in fines just for violations at its oil refineries, while Exxon's fines were less than $400,000.[49]

BP's efforts to expand, gain market share, and become an international leader in energy production provided the organizational environment to pursue production at all costs. BP's corporate culture celebrated risk-taking and embraced the notion of running-to-failure, maintaining a razor-sharp focus on the bottom line. Distinguishing BP's culture from that of its competitors, Stanley Reed and Alison Fitzgerald called it a financial culture rather than one where engineering principles dominated. They noted, "BP is very creative at finding oil and persuading governments to open their doors. But it is sometimes less good at everyday operations."[50]

Insufficient Planning and Preparation

From the moment of the explosion, BP's lack of planning and preparation were evident. The US Bureau of Ocean Energy Management, Regulation, and Enforcement—formerly the Minerals Management Service (more about this agency later in the chapter)—and the Coast Guard issued a biting critique of BP and, to a lesser degree, Transocean in a five-hundred-page report released in 2011. These agencies charged that the loss of life and subsequent pollution "were the result of poor risk management, last-minute changes to plans, failure to observe and respond to critical indicators, inadequate well control response and insufficient emergency bridge response training by companies and individuals responsible for drilling at the Macondo well and for the operation of the *Deepwater Horizon*."[51]

Calling BP's response capacity "underwhelming" and its response plan "embarrassing," the National Commission on the BP *Deepwater Horizon* Oil Spill and Offshore Drilling also lambasted the company.[52] The commission noted that BP had named Peter Lutz as a wildlife expert on whom it would rely—but Lutz had died before BP submitted its plan. BP listed seals and walruses as two species of concern in case of an oil spill in the Gulf—these species never see Gulf waters. And the plan included a link that purportedly was for the Marine Spill Response Corporation website but actually led to a Japanese entertainment site. (To be fair, congressional investigations revealed that the response plans submitted to MMS by ExxonMobil, Chevron, ConocoPhillips, and Shell were almost identical to BP's: they, too, suggested impressive but unrealistic response capacity, and three included the embarrassing reference to walruses.[53])

What seems clear is that BP had a commitment to deepwater drilling but lacked the will to develop appropriate plans for responding to a blowout at the bottom of the sea. Like other oil companies, BP had dealt with cleaning up spills on the water's surface. However, it was woefully ill prepared for the Macondo well explosion a mile below the surface and the enormous amount of oil that spewed from the stricken well as a result. BP CEO Tony Hayward was forced to confess that the company was improvising possible solutions to staunch the oil flow, saying, "There is an enormous amount of learning going on here, because we are doing it for real for the first time."[54]

INEFFECTIVE REGULATORY PRESENCE: DEEP TROUBLE AT THE MINERALS MANAGEMENT SERVICE

It would be nice to lay all blame on the villainous greed of a company that grew too fast and cared too little—behavior that, as noted in Chapter 1, we look to government to thwart through regulations and laws. In this case, however, the agency tasked with protecting the Gulf and its inhabitants, the Minerals Management Service (MMS) within the Department of the Interior, failed miserably. Like BP, the MMS had a history of poor decisions and a dysfunctional organizational culture. Like the *Deepwater Horizon*, the agency hit bottom with the blowout of the Macondo well.

It could be argued that the MMS was set up for failure. President Reagan wanted to make good on his campaign promise to reduce regulations and stimulate economic growth, including expanding domestic oil production—goals shared by Secretary of the Interior James Watt. The MMS was part of that refocusing of government. Created in 1982, the MMS was given competing missions: it had authority to regulate the safety of oil exploration and production, and it was also charged with selling leases and collecting revenues from offshore producers. It quickly became a cash cow for the US Treasury, bringing in billions of dollars from lease sales and royalty payments from oil and gas companies. Revenue generation soon became the agency's central goal, and both government and industry enjoyed seeing increased offshore production. James Watt had promised to lease virtually the entire one-billion-acre Outer Continental Shelf—a staggeringly large offshore area—to oil producers.[55] Revenues paid by oil and gas companies represented the second-largest revenue source for the federal government. However, collecting this bounty led to an all-too-cozy relationship with the companies and an attendant disregard for monitoring safety protocols.

It is worth mentioning that political support for offshore drilling in the Atlantic or Pacific Oceans was far from universal, especially as oil prices dropped in the 1980s and 1990s, making oil and gas seem plentiful. Congress imposed moratoriums on new offshore leases everywhere but in the Gulf and smaller areas in Alaska, a position that was underscored by President George H. W. Bush, who canceled all sales off the California, southern Florida, Washington, and Oregon coasts and in the North Atlantic for ten years (1990–2000).[56] Thus, the oil-rich Gulf became even more attractive to offshore oil producers. This set the stage for both concentrated offshore operations and concentrated political power.

In 2008, the MMS became embroiled in a wide-ranging ethics scandal involving drug use, promiscuity, and graft. Investigations of the agency revealed numerous examples of conflicts of interest and unprofessionalism on the part of MMS staff. In submitting three reports to Congress, the department's inspector general, Earl E. Devaney, noted serious concerns with the integrity and behavior of more than a dozen current and former MMS employees. The reports, spanning two years of investigation, revealed that MMS employees accepted gifts and trips from oil company executives with whom they conducted official business, engaged in sexual encounters with oil and gas company representatives, failed to collect payments due, and manipulated contract negotiations—all direct violations of government ethics rules.

Devaney's cover memo warned that a "culture of ethical failure" pervaded the agency: "[The] single-most serious problem our investigations revealed is a pervasive culture of exclusivity, exempt from the rules that govern all other employees of the federal government."[57] His sharpest criticism was leveled against MMS staff members who operated the "royalty in kind" program, which collected oil and gas in lieu of cash for resale in oil and gas markets.[58] This created a strong motivation for MMS staff to favor promoting oil sales over regulating oil companies. After this blistering indictment by the inspector general, several high-ranking officials in the troubled agency left; one was criminally prosecuted. But that did not end the problems at MMS.

The too-cozy relationship also permeated the agency's rule-making process. During the George W. Bush administration, the MMS developed a hands-off attitude toward regulating oil companies, often deferring to the judgment of oil company officials when deciding whether to inspect or regulate drilling activities. In 2003, the MMS decided against requiring offshore drillers to install an acoustic switch, a remote-controlled backup system to seal an underwater well even if the rig above is destroyed.[59] Such a switch is mandated by Brazil and Norway.

As early as 2004, the MMS knew that blind shear rams—which, as noted earlier, seal off out-of-control oil and gas wells by pinching the pipe closed—might not work well in deepwater drilling operations. Yet at the time of the Macondo well failure, the MMS had yet to act on this information. It also failed to promulgate additional regulations on cementing practices, even in the face of multiple tests indicating that blowouts due to faulty well casings were likely.

As oil gushed from the Gulf of Mexico seafloor, the media honed in on a glaring omission of the MMS: its failure to examine regional and facility oil spill response plans, a requirement of the Clean Water Act. Rubber-stamped by the MMS, BP's plan was a boilerplate document that predicted that no oil from spills would reach the shore, in part because of its overly confident assessment of its skimming operations and the amount of oil that could potentially be leaked from a failed operation.[60] The spill response plan was riddled with these and other errors, but the agency was silent about all of them.

Finally, there was the decision made by the MMS on how to fulfill its obligations under the National Environmental Policy Act (NEPA). NEPA requires that federal agencies, in this case the MMS, conduct a full environmental review of any major federal action significantly affecting the quality of the human environment before that action is undertaken. One exception to this requirement occurs when a federal agency determines that a group, or category, of activities has been found to have no serious individual or cumulative environmental impacts. Thus, federal actions within the category are part of a "categorical exclusion" and do not require either individual environmental assessments, or the more extensive Environmental Impact Statements.

In 2009, the MMS approved BP's exploration plan for the Macondo well under a categorical exclusion, observing that it was "unlikely that an accidental surface or subsurface oil spill would occur from the proposed activities."[61] In granting the lease, the MMS further certified that the company had "the capacity to respond, to the maximum extent practicable, to a worst-case discharge, or a substantial threat of such a discharge."[62] In granting this exclusion, scholarly studies after the blowout suggest that the MMS violated even its own guidance, since BP wanted to drill at a depth that was different from the majority of exploration wells in the Gulf.[63] With MMS thus exempted from conducting a detailed analysis of the environmental impacts of BP's operations, neither the company nor the agency was prepared for the consequences of the *Deepwater Horizon* fiasco.[64] By maintaining a belief that a blowout in deep water was not supposed to happen, despite tests that showed equipment behaved differently in deep water, the MMS fueled the fiction that such drilling would never lead to an environmental catastrophe. But just a year later, it did.

AFTER THE SPILL: BP PAYS UP

The Obama administration, feeling public heat from failing to avert the crisis, pressured BP to meet its financial obligations not only to clean up the Gulf but to compensate the multibillion-dollar tourism and fishing industries and other enterprises affected by the spill. After an intense, four-hour meeting at the White House on June 16, 2010, BP chairman Carl-Henric Svanberg and other BP executives agreed to create a $20 billion claims fund to be administered by an independent claims board.[65] Kenneth Feinberg, the Obama administration official who oversaw compensation for executives at companies that had received federal bailout funds, would oversee the distribution of the fund.

BP made the fund available to satisfy legitimate claims over a four-year period, including natural resource damages and state and local response costs. President Obama stressed that the agreement was only a start: it would not cap BP's total liabilities. The $20 billion figure was roughly what BP had made in annual profits over each of the last four years. Fines and penalties were excluded from the fund and were paid separately by the company. BP chairman Svanberg also offered what might be considered a public apology: "I do thank you for the patience that you have during this difficult time. I hear comments sometimes that large oil companies are greedy companies who don't care. But that is not the case in BP. We care about the small people."[66] To pay for its commitment, BP suspended its dividend and stated that it would sell $10 billion of its assets.

As the oil spilled into the Gulf, the lawyers were preparing claims against BP, which was already engaged in lawsuits against Transocean and

Halliburton. In January 2012, Judge Carl Barbier, who was hearing the lawsuits relating to the disaster in the US District Court for the Eastern District of Louisiana, held that BP was the owner of the well and consequently Transocean's contract with BP shielded it from compensatory claims.[67] Transocean, however, would be subject to any pollution fines levied under the Clean Water Act, as well as its share of punitive damages, if any were assessed.

On May 2, 2012, the same federal district court gave preliminary approval for a settlement compensating victims in two areas: economic and property damages, and medical claims. BP agreed to pay $7.8 billion to satisfy both sets of claims. The court established the *Deepwater Horizon* Court-Supervised Settlement Program to receive claims in this massive class action lawsuit, and a settlement fairness hearing was set for November 8, 2012, to review claims and concerns. But litigation would continue for years. BP also faced suits from other contractors involved in the *Deepwater Horizon* spill and suits filed by people who either were not covered by the agreement or had opted out of the agreement by November 1, 2012. State courts would also be venues for lawsuits against BP.

Even beyond this dizzying array of lawsuits, settlements, and investigations, the federal government was not done with BP. From 2010 to 2012, the Department of Justice continued to bring both criminal and civil prosecutions against BP, drawing on provisions in the Clean Water Act and the Oil Pollution Act. In September 2012, US prosecutors indicated in a filing with the court that they would seek $21 billion in civil damages stemming from gross negligence under the Clean Water Act, crushing BP's hopes of settling for anything less than $15 billion.[68] BP would pay this amount in addition to any settlements reached with victims or governments. A key point of contention in the settlement was the degree to which BP was negligent, because that would determine the scale of the damages.[69]

At least two factors muddied the waters of an early settlement between the federal government and BP. The first was that 2012 was an election year. BP hoped that a new administration might respond more favorably to reducing settlement amounts and finding its actions to be something less than grossly negligent. At the same time, congressional leaders in the Gulf states, in a rare show of bipartisanship, appealed to the Obama administration not to let BP off the hook and to not settle for anything less than the maximum fine. At issue was the amount of money these states would receive under the Resources and Ecosystems Sustainability, Tourist Opportunity, and Revived Economies of the Gulf States Act of 2011 (the RESTORE Act), which directed that 80 percent of the fines levied on BP under the Clean Water

Act go directly to the Gulf states rather than to the federal government. The amount would be based on the number of barrels of oil spilled and the level of negligence.

However, another choice was possible: the federal government could use the Oil Pollution Act to assess any civil or criminal penalties due the government. This law would not deliver fines received back to the Gulf states, but rather direct any monetary fines to the US Treasury. BP preferred settling its criminal and civil penalties under the Oil Pollution Act, as fines under that act would also give the company a more favorable tax treatment, but ultimately the Justice Department prosecuted BP under Clean Water Act provisions.[70]

In January 2013, BP pled guilty to fourteen criminal charges and settled with the Justice Department to pay $4 billion in penalties. Transocean settled civil and criminal claims with the federal government and paid $1.4 billion in penalties. On April 4, 2016, after a finding of gross negligence on the part of BP, Judge Barbier granted final approval of a $20 billion settlement, which included $5.5 billion in civil Clean Water Act penalties. All these fines and penalties were in addition to the trust fund that BP established at the time of the oil spill.

Just how much BP will ultimately pay for ecosystem restoration has not been determined as of this writing. In 2011, the trustees for the *Deepwater Horizon* natural resource damage assessment (NRDA) and BP announced an unprecedented agreement to provide $1 billion for early restoration to address environmental impacts from the oil disaster. But this is just a down payment on the eventual ecological damages, as noted earlier. If one were to use the payment made by Exxon for the environmental costs associated with the *Valdez* oil spill calculated per barrel of oil spilled, BP could pay upwards of $31 billion in restoration costs.[71]

In the first year after the spill, over four thousand miles of coastline were affected, prompting fishing restrictions, beach closures, and the loss of over six thousand birds, six hundred sea turtles, and seven hundred dolphins. Most of the sea turtles were the endangered Kemp's Ridley sea turtle, which lives almost no other place on earth, and its endangered status focused attention on it as an indicator species. As of 2012, approximately 1,145 turtles had been affected, and over half of them succumbed to the effects of the oil.[72]

The time, location, and enormity of the *Deepwater Horizon* catastrophe only compounded the ecological losses. Mississippi Canyon is ecologically sensitive, with richly diverse deepwater sea life. However, the same can be said of the entire Gulf. In its ten-year report, the Census of Marine Life highlighted the Gulf of Mexico as one of the top five most diverse areas of the world in terms of known species.[73] A rare stroke of good luck happened when the Loop Current, which functions like a conveyor belt bringing warm water into the Gulf, destabilized in May 2010, sparing the Florida coast from getting as much oil as originally feared.[74]

The timing was bad for migratory birds that rely on the Gulf region, not to mention the herons, egrets, and brown pelicans that called the Gulf coast home. This was also the time of year when many species were spawning and breeding, so many young were exposed when they were most vulnerable. Perhaps most serious was the damage to the Louisiana wetlands, which represent about 40 percent of the coastal wetlands found in the continental United States.

And the damage may continue. Two years after the spill, scientists continued to examine the health of fish and marine life populations. Scientists estimated that roughly 1.1 million barrels of oil released during the spill were still in the Gulf. NOAA had found hydrocarbons in sufficient quantities along the Louisiana shore to keep areas closed to fishing until April 19, 2011—almost a year to the day after the spill. Underground plumes of oil as long as thirty-five kilometers confirmed that vast quantities of oil did not reach the surface and may cause continued problems for sea life, including the toxic impact of the oil on microscopic plants and the marine animals physically coated with it.[75] Hydrocarbon-eating microbes consume the oil in the water, but also may prompt the formation of low- or no-oxygen dead zones.

Also unanswered are questions about the environmental consequences of using oil dispersants. Concerns about the toxicity of the dispersant COREXIT to aquatic life prompted the EPA to temporarily halt its use by BP. The company was eventually allowed to resume spraying the dispersant on the oil slick, but only because of the lack of safer alternatives. Little is known about the long-term effects of the dispersant, if any, on phytoplankton and other marine life.[76]

When all the lawsuits are settled, the damages, penalties, and fines are paid, and the cleanup is complete, BP has estimated that costs related to the spill will exceed $53 billion.[77] Exxon's fines and settlement for the *Valdez* spill pale in comparison. Exxon eventually settled with victims after the *Valdez* spill for less than $1 billion, after prevailing in lawsuits that had claims totaling $5 billion. The only close rivals to BP in terms of costs are the six major tobacco companies that agreed in 1998 to pay $246 billion in a multistate settlement over twenty-five years for the health-related effects of cigarette smoking, and a host of asbestos-related companies found liable for exposing thousands of people to unsafe levels of asbestos. Many of these companies sought bankruptcy protections in the last two decades but are still on the hook to pay victims compensation. W. R. Grace is one such company, and the focus of Chapter 5.

AFTER THE SPILL: POLITICAL AND POLICY CHANGES

The backdrop to this story involves US energy policy, partisan politics, and our love affair with fossil fuels. Americans have come to expect cheap gas, and plenty of it. In Louisiana, oil is king. Oil platforms in the state produce 30 percent of US oil and account for 400,000 jobs and $70 billion in economic activity. Oil has also long been a growth industry for Texas, where Houston serves as a hub for oil companies and support industries, and home to one of the country's largest ports. The Gulf Coast states have benefited from the economic engine that is Big Oil. This seemingly unwavering support was epitomized by the apology by Representative Joe Barton (R-TX) to BP for the $20 billion "shakedown" by President Obama. (Barton later retracted his apology to BP, but only after a media firestorm and the threat by Republican leaders to pull him off the powerful House Energy Committee.)[78] However, concentrating any industry in one area can imperil an economy. In 2017, Hurricane Harvey's path through Texas hit at the heart of America's energy industry. Houston, home to almost half of America's refining capacity and 20 percent of its oil production, was especially hard hit when floodwaters and high winds damaged roughly half of the area's oil refineries. The consequences of this hurricane, and the likelihood of others to follow in the Gulf, make continued reliance on Big Oil in the Gulf increasingly precarious.

Even as Harvey devastated a large swath of this oil-dependent region, national political leaders, including President Trump, doubled down on their support of the fossil fuel industry. Shrugging off renewed calls to address climate change in the face of this extreme weather event, the Trump administration promised to restore Houston and once again called for opening new offshore areas to drilling. As mentioned earlier, offshore drilling in US waters had been largely restricted to the Gulf of Mexico and some parts of Alaska since the 1989 *Exxon Valdez* disaster. Angry Americans watched the ecological damage wrought by the tanker when its entire payload of crude oil poisoned the pristine area of Prince William Sound, and demanded that politicians guard against future spills. As a result, political appetites for offshore drilling near coastal states diminished. For nearly a decade, the moratorium on offshore drilling seemed politically invulnerable.

FIGURE 3.1: TIMELINE OF THE BP *DEEPWATER HORIZON* OIL SPILL

2005: BP's Texas City, Texas, refinery explodes, killing 15 workers and injuring 180.

2006: A ruptured pipeline owned by BP spills 200,000 gallons of crude oil in Alaska. OSHA fines BP $2.4 million for unsafe conditions at its Ohio refinery.

2007: Tony Hayward becomes BP's chief executive. A BP review panel finds a breakdown in the culture of safety at BP.

2009: OSHA fines BP $87.4 million for willful safety violations at the Texas City refinery—the largest fine in OSHA's history.

January 2010: The *Deepwater Horizon* drilling rig arrives at the Macondo lease site in the Gulf of Mexico. Crew members dub the Macondo well the "well from hell."

April 2010: *Deepwater Horizon* explodes, killing eleven people and injuring seventeen. Oil leaks from the Macondo well. President Obama announces that no new drilling will occur until the cause of the accident is established. BP CEO Tony Hayward promises that BP will take full financial responsibility for the spill. Attempts to contain the flow of oil fail.

May 2010: President Obama announces the creation of the National Commission on the BP *Deepwater Horizon* Oil Spill and Offshore Drilling. The Minerals Management Service is reorganized into three independent entities.

June 2010: The US government begins a criminal investigation into the accident. BP's containment efforts capture ten thousand barrels of oil per day, about one-fourth of the amount spewing from the well. BP agrees to put $20 billion into a fund to compensate victims of the spill and cancels stock dividends for the next three quarters.

July 2010: BP finally succeeds at stopping the oil spill. An estimated 134 million gallons of oil leaked, making this the biggest spill in US history.*

September 2010: US Coast Guard Admiral Thad Allen declares that the Macondo well is effectively dead.

October 2010: Obama administration lifts the ban on deepwater drilling.

2011: The National Commission releases its final report, concluding that the accident could have been prevented.

2012: BP agrees to pay $4 billion in criminal fines; a year later a BP engineer is found guilty of obstruction of justice.

2015: BP agrees to pay $20.8 billion to settle charges with the federal government and five Gulf Coast states for penalties under environmental laws, representing the largest environmental settlement in US history.

2016: Halliburton and Transocean settle punitive damage claims for $1.24 billion, adding to the $1.3 billion paid in 2014.

2017: President Donald Trump signs an executive order directing the Department of the Interior to review safety rules on offshore drilling.

* See note no. 10

"Stop and Think" or "Drill, Baby, Drill"?

By 2001, things began to change. Two men with strong ties to the oil industry, President George W. Bush and Vice President Dick Cheney (former CEO of Halliburton), were now in the White House. At the same time, gas prices were rising. The political winds had shifted, and the oil industry was gaining even more influence in national politics.

Within two weeks after taking office, Bush formed a task force, the National Energy Policy Development Group, to develop suggestions for a new US energy policy, and he named Cheney as its chairman. Cheney's closed-door meetings with executives and lobbyists from the oil, gas, and coal industries angered environmentalists. Not surprisingly, the national energy policy that emerged from the task force encouraged developing additional sources of domestic oil. The terrorist attack on September 11, 2001, opened this window of opportunity even wider to expand drilling in the Gulf as well as elsewhere in the United States. Concerns about energy cost and supply now were joined with concerns about energy security.

By 2008, gas prices exceeded $4 a gallon—a first for Americans. Bush had lifted his father's ban on new drilling, leaving it up to Congress to address its long-standing ban on further offshore drilling. Interest groups, such as the "American Solutions" group headed by former House Speaker Newt Gingrich, began an intensive lobbying campaign. The call to "drill here and drill now" found traction with people who were feeling the pinch of higher gas prices. By mid-2008, more than half of Americans polled backed offshore drilling in areas that were currently off-limits.[79] During the height of the presidential campaign that year, the Republican nominee, Senator John McCain, and his running mate, Alaska governor Sarah Palin, galvanized support through their rallying cry "drill, baby, drill." Although the Democratic nominee, Senator Barack Obama, was initially opposed to expanding drilling in the Outer Continental Shelf, he moderated his position to support a "careful and responsible" approach to new offshore drilling. By the end of September, Congress had lifted the moratorium. Obama handily won the election—and soon made a decision that would come back to haunt him.

As one of his last acts in office, President Bush had issued a plan to open vast areas, from Alaska to the Atlantic coast, for exploration. This act was immediately delayed by the new interior secretary, Ken Salazar, who opened the plan for public comment. However, not only was Salazar under pressure from the American Petroleum Institute, which claimed that such a delay would slow job creation, but he was also besieged by the tumult he inherited at the MMS. Just thirty-three days after Obama's inauguration, BP submitted its request to explore Mississippi Canyon Block 252.

Less than a month before the *Deepwater Horizon* met its fiery demise, President Obama made what turned out to be a grievous political miscalculation: he announced his intention to open part of the American coastline to oil and natural gas drilling, much of it for the first time. This apparent reversal of his reluctance to expand offshore drilling was part of a vote-trading effort: by expanding drilling off the Southeast coast, he hoped to win political support for energy and climate legislation from otherwise recalcitrant Republicans. Just three weeks later, the Macondo well blew, and the reputation of an administration that had stood against the notion of "drill, baby, drill" was damaged in the aftermath.

Punctuated Equilibrium: New Policies on Offshore Drilling and Agency Reorganization

As described in Chapter 1, punctuated equilibrium theory helps explain why public policies that remain stable for long periods of time can undergo rapid and substantial change. Policies stay the same owing to "bounded rationality" (that is, policymakers' inability to consider all problems); lack of attention; the way an issue is framed; and policy monopolies—certain groups defining the policy problem and the solution for a long period of time. A policy may be changed when public attention is drawn to its shortcomings and new solutions are demanded. The BP story seems to be one of punctuated equilibrium—policy changes occurring after decades of quiescence.

Extensive media coverage—especially the live television feed of gushing oil and the images of dead sea turtles and dolphins—provided the catalyst to make policy changes and right the course of the MMS and federal regulatory oversight of deepwater drillers. On May 28, 2010, President Obama imposed a six-month moratorium on deepwater drilling in order to assess the safety of existing practices on rigs. This came roughly one week after Representative Edward Markey (D-MA) revealed during congressional hearings in the aftermath of the BP spill that the oil spill response plans of the major oil companies were virtually identical. Even worse, these plans, supposedly scrutinized by the MMS, protected the cold-water walrus and other species not found in the hot Gulf of Mexico—clearly indicating that no company, not just BP, had fully developed a plan for its deepwater drilling operation. Caught flat-footed, oil executives were forced to admit that they had no serious plan for dealing with a major oil spill.

In the give-and-take of politics, President Obama's temporary moratorium was overturned by US District Judge Martin Feldman on June 22,
who found insufficient support to shutter the very business on which many
Louisianans depended. In issuing an injunction, he wrote:

> What seems clear is that the federal government has been pressed by
> what happened on the *Deepwater Horizon* into an otherwise sweeping
> confirmation that all Gulf deep water drilling activities put us all in
> a universal threat of irreparable harm. . . . The blanket moratorium,
> with no parameters, seems to assume that because one rig failed and
> although no one yet fully knows why, all companies and rigs drilling
> new wells over 500 feet also universally present an immediate harm.[80]

Frustrated by the court's action, the Obama administration sharpened
the target of the moratorium to the performance of a rig's blowout preventers and other safety protocols, and then it once again imposed a moratorium, on July 12, 2010. This time, if rig operators could prove that the
blowout preventers would quickly shut down an out-of-control well and that
their oil response plans were adequate, they would be allowed to drill. The
Obama administration lifted the moratorium in October for operators who
complied with the new regulations. In announcing that the moratorium was
lifted, Interior Secretary Ken Salazar noted that the new regulatory oversight had significantly reduced the risks of deepwater drilling.[81]

In December 2010, the administration backed further away from its
ill-timed decision to expand offshore oil exploration. Drilling would remain
under a moratorium for areas in the eastern Gulf and along the Atlantic
coast for at least the next seven years, until stronger safety and environmental standards were in place. But drilling would continue in the central and
western Gulf of Mexico, although with a new set of safeguards for drilling,
blowout preventers, worker safety, and response plans. In 2011, the Obama
administration proposed a five-year plan for offshore oil drilling that called
for opening new areas in the Gulf of Mexico and Alaska but barred development along the East and West Coasts.

Policy change arising from punctuated equilibrium can affect the agencies tasked with implementation. The Obama administration wasted no
time in dealing with the MMS, the much-maligned federal agency responsible for policing offshore drilling. On May 19, 2010, Interior Secretary Ken
Salazar signed a Secretarial Order that the MMS be divided into three separate organizations: the Bureau of Safety and Environmental Enforcement

(BSEE), the Bureau of Ocean Energy Management (BOEM), and the Office of Natural Resources Revenue (ONRR), each with a separate and clearly defined mission. MMS was temporarily renamed the Bureau of Ocean Energy Management Regulation and Enforcement (BOEMRE) and given a new director in June 2010. The reorganization separating the functions of the MSS was completed on October 1, 2011.[82] On its website, the Department of the Interior acknowledged that the MMS "could not keep pace with the challenges of overseeing industry operating in US waters."[83]

In addition to dissolving the MMS and splitting it up by function, the Obama administration made sweeping changes to focus the environmental efforts of the agencies. The position of chief environmental officer was created for the first time. Environmental reviews were separated from leasing in the BOEM, and a new environmental compliance and enforcement function was given to the BSEE, along with more prominent responsibility for oil spill response plan review.

The Council on Environmental Quality (CEQ), the agency that oversees NEPA, recommended that the Department of the Interior stop granting categorical exclusions and require a full environmental review of deepwater oil exploration prior to permitting the activity. As noted earlier, the MMS had previously issued dozens of these exemptions to oil companies, including BP. The Interior Department, which had fostered the practice, now undertook a review of all exemptions in the aftermath of the Macondo well blowout, in yet another change to long-standing policy.

CONCLUSION

It is indeed ironic that the *Deepwater Horizon* sank on the fortieth anniversary of the first Earth Day—a day inspired in part by the blowout of a well off the shores of Santa Barbara, California, in 1969. That anniversary reminds us of what can be accomplished when the environment occupies center stage in the American psyche.

Was BP an environmental villain? Perhaps that should be left up to the reader to decide, but it is certainly the case that connections can be drawn between BP's actions and the definition of villainous behavior offered in Chapter 1. Testimony during congressional hearings and the investigations carried out by the presidential commission, the Coast Guard, and BP itself reveals that employees were well aware that the culture was one of "run to failure," despite proclamations to the contrary by the company's senior management. Little doubt remains that BP was not a high-reliability organization, given its history of oil spills, safety violations, and site-specific debarment by the EPA. All this happened before the Macondo well blew.

One cannot forget the grievous refinery accident in Texas in 2005 that killed fifteen people and injured nearly two hundred. According to the "atrophy of vigilance" theory described in Chapter 1, BP should have been motivated to renew its vigilance after the refinery explosion and redouble its efforts to tighten safety practices. The company should have tightened its environmental standards after the oil spill in Alaska dumped two hundred thousand barrels of oil in 2006. Senior management might have listened to the crew who dubbed Macondo the "well from hell," or the engineers who suggested that additional centralizers were needed, as well as another blind shear ram, or those who worried about the failed negative pressure tests after injecting the drilling cement. BP did none of these things.

As a company, BP emerged bruised and battered from its experience with the *Deepwater Horizon*. The company replaced CEO Tony Hayward after the public relations debacle surrounding his seemingly uncaring comment that he "wanted to get his life back" even while oil gushed into the Gulf. This change of leadership seemed to be a pattern too: Tony Hayward had replaced CEO John Browne after the Alaska pipeline spill and massive accident at the Texas refinery.

Robert Dudley, who officially assumed the reins of the company in October 2010, has watched BP sell off a number of its holdings. In September 2012, the company sold its stakes in some Gulf oil fields to Plains Exploration & Production for $5.6 billion.[84] A month later, BP announced the sale of its Texas City refinery to rival Marathon Oil for $2.5 billion, bringing the total asset sales since the oil spill to about $35 billion. Though more streamlined, BP intends to keep drilling in the Gulf.

The government, too, has changed as a result of the oil spill. When the MMS was dissolved and its functions divided among three separate agencies, two of them were newly created as a consequence of the BP oil disaster. Tougher new regulations are in place now for offshore drilling. New environmental reviews under NEPA are now part of the process, and environmental groups are reenergized in their efforts to prevent new leases for offshore drilling for oil.

However, President Donald Trump has signaled his intention to expand offshore drilling for oil and gas. In a dramatic departure from the previous administration, the president signed an Executive Order in 2017 directing the Secretary of the Interior, Ryan Zinke, to review a five-year plan of the Obama administration that banned drilling in parts of the Pacific, Arctic, and Atlantic Oceans. Opening new areas for offshore drilling is likely to meet with resistance from the affected communities and from environmental groups.

Perhaps the best example of changing politics is drilling in the Arctic. One of President Obama's last acts before leaving office was to forbid oil and gas drilling in nearly all US waters in the Arctic, employing powers granted in the Outer Continental Shelf Lands Act to "withdraw from disposition" of unleased lands. The ban protects about 115 million acres of federally owned Arctic waters, as well as nearly 4 million acres of coral canyons off the Atlantic Coast.[85] The move was one of several attempts by President Obama to protect the environment in the waning days of his presidency.

Changes such as the major reorganization of the MMS, tougher safety and environmental reviews, and new rules requiring the testing of blowout preventers are promising. So, too, is the RESTORE Act, which will direct 80 percent of the fines levied against BP under the Clean Water Act to the Gulf states in an ongoing effort to make the Gulf ecosystem whole. Whether or not this is enough to prevent another catastrophe remains to be seen. Deepwater drilling continues, and political winds change. With the Trump administration's commitment to rolling back regulations, the winds are likely to favor oil and gas development for the foreseeable future.

What about the rest of us? Do Americans have the willpower to wean themselves off fossil fuels and shift to renewable sources of energy? Considering how little appetite we have shown for doing so in the past, this remains an open question. After all, Sarah Palin did not chant "drill, baby, drill" by herself. Thousands joined in. During the election campaign of 2012, the Republican nominee, Mitt Romney, promised that he would open additional public lands and federal waters to drilling for oil, a promise repeated by Donald Trump during his campaign and now during his administration. Big Oil is big business and an economic powerhouse. It is also a business that does great environmental harm, whether from the methods we use to take oil from the earth or from burning it and creating greenhouse gases that contribute to climate change. We have established the conditions under which BP and other oil companies can thrive by demanding more oil, more domestic production, bigger cars, and cheaper gas. It's up to us to decide where we go from here.

DISCUSSION QUESTIONS

1. What technical, political, and organizational factors account for the sinking of the *Deepwater Horizon*?
2. How did BP's culture influence its environmental and safety practices?

3. Why was the MMS so ineffective in regulating offshore drilling?

4. How can we influence the direction of US energy policy?

NOTES

1. William Freudenburg and Robert Gramling, *Blowout in the Gulf: The BP Oil Spill Disaster and the Future of Energy in America* (Cambridge, MA: MIT Press, 2011).

2. BP, "BP Announces Giant Oil Discovery in the Gulf of Mexico" (press release), September 3, 2009, www.oilandgasonline.com/doc/bp-announces-giant-oil-discovery-in-the-gulf-0001.

3. Ibid.

4. Transocean, "Our Company: Global Operations," 2012, http://www.deepwater m/fw/main/Global-Operations-8.html (accessed September 20, 2012).

5. National Commission on the BP *Deepwater Horizon* Oil Spill and Offshore Drilling, *Deep Water: The Gulf Oil Disaster and the Future of Offshore Drilling: Report to the President* (Washington, DC: US Government Printing Office, January 2011), ix, www.gpo.gov/fdsys/pkg/GPO-OILCOMMISSION/pdf/GPO-OILCOMMISSION.pdf (accessed June 7, 2017).

6. Bob Cavnar, *Disaster on the Horizon: High Stakes, High Risks, and the Story Behind the Deepwater Well Blowout* (White River Junction, VT: Chelsea Green Publishing, 2010).

7. BP, *Deepwater Horizon* Accident Investigation Report (internal document), September 8, 2010, 29, www.bp.com/content/dam/bp/pdf/sustainability/issue-reports/Deepwater_Horizon_Accident_Investigation_Report_Executive_summary.pdf.

8. Peter Lehner and Bob Deans, *In Deep Water: The Anatomy of Disaster, the Fate of the Gulf, and Ending Our Oil Addiction* (New York: The Experiment LLC, 2010), 1.

9. National Commission on the BP *Deepwater Horizon* Oil Spill and Offshore Drilling, *Deep Water*.

10. The total amount of oil spilled will never be known. The federal government estimated the amount at 4.2 million barrels; BP estimated 2.4 million barrels. Both took into account the amount of oil recovered during cleanup. For purposes of calculating penalties under the Clean Water Act, Judge Carl Barbier of the US District Court for the Eastern District of Louisiana ruled that 3.19 million barrels, or about 134 million gallons, had been spilled. This is the figure used in the book.

11. Cavnar, *Disaster on the Horizon*, 25.

12. Bureau of Ocean Energy Management, Regulation and Enforcement, *Report Regarding the Causes of the April 20, 2010 Macondo Well Blowout* (Washington, DC: US Department of the Interior, September 14, 2011 https://www.bsee.gov/sites/bsee.gov/files/reports/blowout-prevention/dwhfinaldoi-volumeii.pdf (accessed December 21, 2017).

13. "Timeline: Oil Spill in the Gulf," *CNN*, 2010, www.cnn.com/2010/US/05/03 /timeline.gulf.spill/index.html (accessed October 8, 2012).

14. US Coast Guard, *Final Action Memorandum—Incident Specific Prepared-ness Review (ISPR) Deepwater Horizon Oil Spill* (Washington, DC: Department of Homeland Security, March 18, 2011), www.uscg.mil/foia/docs/DWH/BPDWH.pdf (accessed June 8, 2017).

15. Marcia K. McNutt, Rich Camilli, Timothy J. Crone, George D. Guthrie, Paul A. Hsieh, Thomas B. Ryerson, Omer Savas, and Frank Shaffer, "Review of Flow Rate Estimates of the *Deepwater Horizon* Oil Spill," *Proceedings of the National Academy of Sciences of the United States of America* 109, no. 50 (2012): 20260–20267.

16. Ibid.

17. Campbell Robertson and Henry Fountain, "BP Says Oil Flow Has Stopped as Cap Is Tested," *New York Times,* July 15, 2010.

18. *CNN,* "Timeline: Oil Spill in the Gulf."

19. Cavnar, *Disaster on the Horizon,* 122.

20. David Biello, "How Science Stopped BP's Gulf of Mexico Oil Spill," *Scientific American* (April 19, 2011). https://www.scientificamerican.com/article/how-science -stopped-bp-gulf-of-mexico-oil-spill/ (accessed November 1, 2017).

21. *CNN,* "Timeline: Oil Spill in the Gulf."

22. Abraham Lustgarten, "Furious Growth and Cost Cuts Led to BP Accidents Past and Present," *ProPublica,* October 26, 2010, www.propublica.org/article/bp -accidents-past-and-present (accessed October 16, 2012).

23. David Hammer, "BP Was More Than $40 Million over Budget for Blown-out Well, Oil Spill Hearings Show," *(New Orleans) Times Picayune,* August 26, 2010.

24. National Commission on the BP *Deepwater Horizon* Oil Spill and Offshore Drilling, *Deep Water,* 4.

25. Cavnar, *Disaster on the Horizon,* 43.

26. Lehner and Deans, *In Deep Water,* 15; Abraham Lustgarten, *Run to Failure: BP and the Making of the Deepwater Horizon Disaster* (New York: W. W. Norton and Co., 2012), 308.

27. Robbie Brown, "Official Denies BP Put Cost Ahead of Safety at Oil Rig," *New York Times,* July 23, 2010.

28. Lustgarten, *Run to Failure,* 310.

29. National Commission on the BP *Deepwater Horizon* Oil Spill and Offshore Drilling, *Deep Water,* 97.

30. Cavnar, *Disaster on the Horizon,* 27.

31. Emmett Mayer, "Six Steps That Doomed BP's *Deepwater Horizon* Oil Rig," *(New Orleans) Times-Picayune,* September 5, 2010.

32. David Barstow, Laura Dodd, James Glanz, Stephanie Saul, and Ian Urbina, "Regulators Failed to Address Risks in Oil Rig Fail-Safe Device," *New York Times,* June 20, 2010.

33. Robbie Brown, "Oil Rig's Siren Was Kept Silent, Technician Says," *New York Times,* July 24, 2010.

34. Ibid.

35. Ibid.

36. Loren C. Steffy, *Drowning in Oil: BP and the Reckless Pursuit of Profit* (New York: McGraw-Hill, 2011), xvi.

37. Freudenburg and Gramling, *Blowout in the Gulf*, 40.

38. Tom Price and T. J. Aulds, "What Went Wrong: Oil Refinery Disaster," *Popular Mechanics*, March 23, 2005. www.popularmechanics.com/technology/gadgets/2295/1758242/ (accessed November 1, 2017).

39. US Chemical Safety and Hazard Investigation Board (CSB), "Investigation Report: Refinery Explosion and Fire," Report 2005-04-I-TX, March 23, 2005.

40. Josh Cable, "Anatomy of a Tragedy," *Occupational Hazards* (October 2008): 41.

41. Ibid.

42. CSB, "Investigation Report: Refinery Explosion and Fire," 20.

43. Lustgarten, "Furious Growth and Cost Cuts Led to BP Accidents Past and Present."

44. Cable, "Anatomy of a Tragedy," 41.

45. CSB, "Investigation Report: Refinery Explosion and Fire," 20. A willful violation describes a voluntary action that intentionally disregards a regulatory requirement.

46. Lustgarten, "Furious Growth and Cost Cuts Led to BP Accidents Past and Present."

47. Ibid.

48. Ibid.

49. Ibid.

50. Stanley Reed and Alison Fitzgerald, *In Too Deep: BP and the Drilling Race That Took It Down* (Hoboken, NJ: Bloomberg Press, 2011), xii.

51. Bureau of Ocean Energy Management, Regulation and Enforcement, *Report Regarding the Causes of the April 20, 2010 Macondo Well Blowout* (Washington, DC: US Department of the Interior, September 14, 2011), 1, https://www.bsee.gov/sites/bsee.gov/files/reports/blowout-prevention/dwhfinaldoi-volumeii.pdf (accessed December 21, 2017).

52. National Commission on the BP *Deepwater Horizon* Oil Spill and Offshore Drilling, *Deep Water*, 133.

53. Ibid.

54. Quoted in Reed and Fitzgerald, *In Too Deep*, 70.

55. National Commission on the BP *Deepwater Horizon* Oil Spill and Offshore Drilling, *Deep Water*, 63.

56. Ibid., 67.

57. Earl Devaney, Office of the Inspector General, US Department of the Interior, memorandum to Secretary Kempthorne, "Subject: OIG Investigations of MMS Employees," September 9, 2008, http://s3.amazonaws.com/propublica/assets/docs/oig_devaney_letter_080909.pdf (accessed October 31, 2012).

58. Charlie Savage, "Sex, Drug Use, and Graft Cited in Interior Department," *New York Times*, September 11, 2008.

59. Kate Sheppard, "There Will Be Blood," *Mother Jones,* May 5, 2010, www.mother jones.com/politics/2010/05/bp-bill-nelson-oil-spill (accessed October 31, 2012).

60. Thomas A. Birkland and Sarah E. DeYoungy, "Emergency Response, Doctrinal Confusion, and Federalism in the *Deepwater Horizon* Oil Spill," *Publius: The Journal of Federalism* 41, no. 3 (2011): 471–493, 474.

61. Sheppard, "There Will Be Blood."

62. Ibid.

63. Maria Luwalhati Dorotan, Marija Janes, and Ana Olaya, "Categorical Exclusion on Deepwater Offshore Drilling: Before and After the BP Blowout," Tulane Law School (undated), http://www.law.tulane.edu/uploadedFiles/Academics/Lectures /Categoricl%20Exclusion%20On%20Deepwater%20Offshore%20Drilling%20.pdf (accessed September 21, 2017).

64. Juliet Eilperin, "US Exempted BP's Gulf of Mexico Drilling from Environmental Impact Study," *Washington Post,* May 5, 2010.

65. BP, "BP Establishes $20 Billion Claims Fund for *Deepwater Horizon* Spill and Outlines Dividend Decisions" (press release), June 16, 2010, http://www.bp.com /genericarticle.do?categoryId=2012968&contentId=7062966 (accessed September 15, 2012).

66. Jeff Mason, "BP Agrees to $20 Billion Spill Fund, Cuts Dividend," *Reuters,* June 16, 2010.

67. John Broder, "Transocean Not Liable for Some Gulf Spill Claims, Judge Rules," *New York Times,* January 26, 2012.

68. Sarah Young and Sinead Cruise, "DOJ Language Crushes BP Oil Spill Settlement Hopes," *Reuters,* September 5, 2012.

69. Daniel Gilbert, "BP Close to Spill Settlement: Multibillion-Dollar Deal with US Would Combine Civil, Criminal Liabilities," *Wall Street Journal,* October 10, 2012.

70. Roberta Rampton, "BP Oil Spill Settlement: Gulf Coast Senators Sign Bipartisan Letter Seeking Fair Deal," *Reuters,* October 5, 2012.

71. Associated Press, "Gulf Restoration After Oil Spill Should Include Conservation Land Purchases, Environmental Coalition Reports," *(New Orleans) Times Picayune,* July 18, 2012.

72. Ocean Conservancy, "BP *Deepwater Horizon* Oil Disaster: Impacts and Studies: Known and Published Impacts," https://oceanconservancy.org/wp-content /uploads/2017/04/bp-deepwater-horizon-oil.pdf.

73. Census of Marine Life, "What Lives in the Sea? Census of Marine Life Publishes Historic Roll Call of Species in 25 Key Ocean Areas" (press release), August 2, 2010, www.coml.org/http%3A/%252Fwww.coml.org/press-releases-2010 (accessed October 10, 2012).

74. Paul Voosen, "Loop Current Destabilizes, Lowering Gulf Oil Spill's Threat to Fla.—for Now," *New York Times,* May 20, 2010.

75. David Biello, "Massive Oil Plume Confirmed in the Gulf of Mexico," *Scientific American,* August 19, 2010. https://www.scientificamerican.com/article/massive -oil-plume-confirmed-in-gulf-of-mexico/ (accessed November 1, 2017).

76. David Biello, "Is Using Dispersants on the BP Gulf Oil Spill Fighting Pollution with Pollution?" *Scientific American,* June 18, 2010. https://www.scientific american.com/article/is-using-dispersants-fighting-pollution-with-pollution/ (accessed November 1, 2017).

77. Associated Press, "Judge Approves $20 Billion Settlement in BP Oil Spill," *New York Times,* April 4, 2016.

78. Jackie Calmes, "Republican Backpedals from Apology to BP," *New York Times,* June 17, 2010.

79. Reed and Fitzgerald, *In Too Deep,* 86.

80. Bryan Walsh, "Obama's Drilling Moratorium Is Moratoriumed," *Time,* June 22, 2010.

81. Tom Cohen, "Obama Administration Lifts Deep-Water Drilling Moratorium," *CNN,* October 12, 2010, http://www.cnn.com/2010/US/10/12/drilling .moratorium/index.html (accessed September 21, 2017).

82. US Department of the Interior, "Interior Department Completes Reorganization of the Former MMS" (press release), September 30, 2011, www.doi.gov/news /pressreleases/Interior-Department-Completes-Reorganization-of-the-Former -MMS.cfm (accessed November 1, 2012).

83. US Department of the Interior, Bureau of Ocean Energy Management, "The Reorganization of the Former MMS," www.boem.gov/About-BOEM /Reorganization/Reorganization.aspx (accessed November 1, 2012).

84. Steven Mufson, "BP Sells Some of Its Gulf of Mexico Assets for $5.6 Billion," *Washington Post,* September 10, 2010.

85. Coral Davenport, "Obama Bans Drilling in Parts of the Atlantic and the Arctic," *New York Times,* December 20, 2016.

CHAPTER 4

Appalachian Coal Country

EXPLOSION AT THE UPPER BIG BRANCH MINE

On April 5, 2010, just fifteen days before the *Deepwater Horizon* explosion, another explosion occurred. This enormously powerful blast killed twenty-nine coal miners, making it the deadliest mine disaster in the United States in over forty years. Like the 1984 tragedy in Bhopal and the sinking of the *Deepwater Horizon,* the explosion at the Upper Big Branch (UBB) mine could be called an accident waiting to happen.

Miners had been pressed to do more by the mine's operator, Performance Coal Company, and its parent company, Massey Energy. Safety protocols were overlooked, and supervisors, enmeshed in a corporate culture that seemingly had little regard for workers or the environment, ignored the warning signs of imminent danger. When the mine blew, the explosion not only took the lives of twenty-nine men and seriously injured another; it closed the mine and shut down a corporate coal mining giant. In the aftermath, the disaster forced the retirement and subsequent criminal prosecution of its larger-than-life chairman and CEO, Donald Blankenship, and prompted extensive federal and state investigations of coal mining practices in West Virginia.

In many ways, however, this story is bigger than the mine disaster, tragic as it was. It is also about the economic injustices that have long prevailed in Appalachian coal country, where coal companies have extracted resources worth millions and where some of the poorest people in America live and always have. This gripping poverty is not likely to end soon. Coal continues to predominate as a worldwide source of cheap energy, but this will not always be the case—the push for cleaner energy is on the rise both nationally and globally. Utility companies across the country are opting to retire coal-fired power plants and burn cheaper natural gas. The deep economic reliance of many Appalachian communities on the coal industry makes those communities especially vulnerable to declining demand for coal. Thus, the poverty of the people in Appalachia, their fate tied to the fate of King Coal, persists.

The environmental costs of coal production are high as well, as described in the next section. Mountaintop removal practices reshape the mountains into something resembling moon craters, pushing vast amounts of "overburden" (the soil, trees, and rocks that overlay coal seams) into the valleys below, threatening communities, and contaminating water supplies. Different problems are created by underground "longwall" mining: the massive shearing of coal seams deep beneath the surface creates subsidence that disturbs the landscape and affects both surface water and groundwater. Moreover, burning coal has a serious impact on the environment. Coal-fired power plants are carbon-intensive and the nation's top source of carbon dioxide (CO_2) emissions, a major contributor to global warming.

Like the disasters described in the previous chapters, coal production represents both a human tragedy and an ongoing environmental concern. Moreover, the story of the Upper Big Branch mine explosion calls us to consider the extent to which our desire for readily available fossil fuel energy feeds into the likelihood that more environmental damage will occur and that more lives will be lost.

This chapter begins by briefly describing the history of coal as a prevalent fuel source, the environmental consequences of mining and burning coal, the coal regions in the United States, mining methods, laws protecting miners, and the changing dynamics of coal as a source of electricity generation. It explores the factors leading up to the tragedy at the Upper Big Branch mine and compares those factors to the common characteristics of industrial disasters discussed in Chapter 1. The chapter ends with a discussion of the court cases, policy responses, and regulatory changes in the aftermath of the explosion of the UBB mine.

A PRIMER ON US COAL

Coal is mined in twenty-four states located in three coal regions of the United States: the West, the Interior, and Appalachia. Five states (Wyoming,

West Virginia, Kentucky, Pennsylvania, and Illinois) accounted for about 71 percent of all coal produced in 2015.[1] Not all coal regions are equal. The vast coal reserves found in the West create economies of scale that allow mine operators to extract coal cheaply. Although the Appalachian region produces about 25 percent of US coal, cheaper coal production in the West has put pressure on mining companies in other regions to cut costs. Mountaintop removal was one such cost-cutting measure; longwall mining was another. Appalachian coal companies are struggling, however, because even the use of these techniques has not challenged the coal production dominance of the Western region.

West Virginia, the second-largest coal-producing state behind Wyoming, dominates coal production in Appalachia, with over 95 million short tons of coal taken from 151 mines in 2015.[2] However, coal production in the Appalachian region has fallen by 53 percent, more than the 36 percent decline in the West. Over 100 Appalachian mines closed between 2014 and 2015, as the total went from 804 mines to 694. The Interior region, which supplies less than 20 percent of US coal, slightly increased production, by 2 percent, from 2008 to 2016, even though 12 mines closed.[3] Whether the gradual decline in US coal production continues will be mostly based on worldwide and national demand for coal-fired power plants. On the one hand, shifts to natural gas and to renewables are affecting demand for coal; on the other, attempts by the Trump administration and other politicians to shore up the coal mining industry have sparked some short-term interest in expanding coal production, as described in the next section.

Coal: An Abundant Energy Source

The United States is not likely to run out of coal anytime soon. According to the US Energy Information Administration (EIA), the country has the largest estimated recoverable coal reserves on the planet, with enough coal to last more than two hundred years. The United States produced nearly one billion short tons of coal in 2015, making it the second-largest coal producer in the world.[4] (The largest coal producer is China, whose coal production has increased over 400 percent since 1980, accounting for 4.2 billion short tons.[5])

Nearly all coal mined in the United States is sent to power plants to generate electricity. The electric power sector accounted for over 92 percent of total coal consumed in 2016.[6] For decades, coal was the dominant source of electricity generation, supplying more than three-fourths of the country's electricity. In recent years, however, power plants have increasingly turned to lower-priced natural gas and renewable sources of energy, such as wind and solar, as part of their energy mix. In 2016, electricity generated from natural gas surpassed electricity generated by coal for the first time in US history, and coal production was at its lowest level since 1978.[7]

Despite competition from natural gas and renewable sources of energy, coal-fired power plants remain important, accounting for 30 percent of the total net electricity produced in 2016.[8] The United States is also a net exporter of coal, and global demand for US coal continues. Exports of coal increased dramatically between 2009 and 2012—to a record high of 125.7 million short tons in 2012, or 12 percent of total global coal production. Exports declined in 2016, with 60.3 million short tons of coal exported, but EIA estimated that coal exports would be somewhat higher (72 million short tons) in 2017.[9]

Unless new and stringent policies to limit emissions of greenhouse gases are put into place, which seems unlikely during the tenure of the coal-friendly Trump administration, the EIA estimates that the amount of coal produced will move roughly in tandem with our demand for power, with some short-term increases in coal production expected in the Western region through 2018.[10] The EIA forecasted in 2013 that US coal production would modestly increase by 0.6 percent every year through 2040.[11] Since that time, however, demand for coal-generated electricity has waned, both in the United States and around the world. While the EIA projects that world energy consumption will increase by 28 percent between 2015 and 2040, global demand for coal is predicted to lose ground to natural gas and renewable sources of energy.[12]

US coal consumption fluctuates with the number of coal-fired power plants in operation. Nearly 150 coal-fired power plants closed in 2015 and 2016, and more closures are planned as these aging plants become too costly to repair (89 percent of the country's plants were built before 1980) and politically unpopular.[13] To put it another way, dramatic increases in demand for coal are unlikely, but the United States will nonetheless be burning coal for many years to come. How many years will be determined in part by the costs of mining coal compared to other energy sources and by political pressures demanding a shift to cleaner and renewable forms of energy. Most of the political pressure demanding a shift away from coal is due to the high environmental costs of coal, as described in the next section.

Coal Mining Methods and Environmental Impacts

Coal is not good news for the environment, as both coal mining and coal burning have adverse consequences for the air, land, and water. When burned, coal emits carbon dioxide (CO_2), sulfur dioxide, nitrogen oxides, mercury, particulates, and heavy metals such as arsenic, contributing to acid rain, smog, and health conditions such as asthma. Perhaps most concerning is coal's contribution to global warming. In 2016, CO_2 emissions, a major greenhouse gas, by the US electric power sector were 1,821 million metric tons, or about 35 percent of total US energy-related CO_2 emissions. Coal accounted for over 68 percent of those emissions.[14]

Politicians have taken very different approaches to the coal industry. In President Obama's view, addressing coal-fired power plants was an essential element in fighting global warming. In response, the EPA released the Clean Power Plan in 2015 to cut carbon emissions from coal plants. The goal of the plan was to cut carbon emissions from coal-fired power plants by 32 percent below 2005 levels by 2030. Achieving this reduction would be up to the states, and some states, as well as the coal industry, sued the EPA over the Clean Power Plan. Other states, such as California, faced off on the other side, wanting swift action on climate change while moving to increase renewable energy on their own. Implementation of the plan was subsequently stayed by the US Supreme Court until lower courts could rule on the legality of the rule.

President Trump and Vice President Mike Pence signaled a dramatic turnaround in US coal policy by vowing to stop the "war on coal." In March 2017, President Trump signed an Executive Order rolling back several climate-related regulations promulgated during the Obama administration. The following month, he signed an Executive Order calling for a review of the Clean Power Plan, with an eye toward dismantling it. He also ordered the relaxation of other environmental regulations that had proved costly to the coal industry and stated his intention to pull out of the international climate agreement, the Paris Accord, described in Chapter 7. Undoubtedly, the national-level politics of coal has tilted once again toward favoring the coal industry. However, countervailing market forces, including the use of natural gas, the cost of upgrading coal-fired power plants, and state and international shifts away from fossil fuels, will temper any quick or permanent upswing in coal demand.

Mining coal also has environmental consequences, which differ depending on the mining technique and the size of the mine. There are two types of mining operations: surface mining and underground mining. Surface mining, as the term implies, involves removing overburden to mine the coal. The overburden is then replaced and the land restored to its "approximate original contour," as required by the Surface Mining Control and Reclamation

Act of 1977. There are gigantic strip mining operations in the West, primarily in the Powder River Basin in Wyoming. Strip mining disturbs thousands of acres of land and disrupts groundwater supplies. Coal mining in the Powder River Basin is almost exclusively done on federal public lands, through coal leases issued by the Bureau of Land Management (BLM) within the US Department of the Interior. In a similar fashion to what happened to offshore oil leases in the Gulf, both the rate at which leases are issued and the amount of regulatory oversight are subject to the policy preferences of the secretary of the interior and the presidential administration.

Perhaps the most egregious form of surface mining to environmentalists is mountaintop removal, a mining practice in Appalachia that does exactly what the term implies—vast quantities of soil are removed through explosions and gigantic drag lines on the mountaintop to extract the coal underneath. Overburden is then deposited into adjacent valleys around the mountain in a process called "valley fill." Mountaintop removal and the dumping of waste and debris is the greatest earth-moving activity in the country, encompassing more than 1.1 million acres—or a total area about the size of the state of Delaware.[15] The EPA estimated that mountaintop removal projects have seriously damaged or buried more than twelve hundred miles of mountain streams.[16]

The ugly scar left on the land by the destruction of over five hundred mountaintops and the clear-cutting of forests has served as a lightning rod for environmental groups and local communities in Appalachia. It was the practice of mountaintop removal that stirred Judy Bonds—whose story is told in Chapter 6—to stand up to Massey Energy.

In underground coal mining, two methods are used: room-and-pillar and longwall. Longwall mining produces coal more efficiently than the old room-and-pillar mining method, in which coal was excavated in small areas, or rooms, around pillars that supported the mining activity. This older technique often required that miners use picks and shovels, and the process was slow and labor-intensive.

In contrast, the longwall method employs continuous mining machines that shear seams of coal deep underground. As the massive cutting head of the machine moves back and forth through the seam, water sprays against the coal surface to dampen the coal dust. The cut coal is moved to the surface on conveyor belts. As the longwall advances through panels of coal, the area behind the shields—which are mounted on hydraulic roof supports—collapses in what the coal industry refers to as "planned subsidence." Once a panel of coal is mined out, the machine is moved to the next panel and the process begins again. Over time, mines employing this technique can become immense labyrinths of underground passages. To get to the equipment in these mines, which can be miles long and more than a mile deep beneath the surface, miners must travel in slow-moving vehicles called "mantrips." The dangers of this technique are many: imagine working in tight confines beside colossal machines a mile underground, often standing knee-deep in water, all while dependent upon fans to provide sufficient oxygen and ventilation.

This country's deep underground longwall mining operations are principally in Appalachia. The Upper Big Branch was such a mine.

The Laws Protecting Miners

The Mine Safety and Health Administration (MSHA) regulates coal mining practices under the authority of the Federal Coal Mine Health and Safety Act of 1969 (also known as the Coal Mine Act). This law was born out of another tragic coal mining explosion in West Virginia in 1968, when seventy-eight miners were killed at the Farmington Number 9 mine. The new law instituted comprehensive occupational safety and health protections, setting standards for coal dust exposure and for black lung benefits and also setting mandatory safety standards for ventilation and rock dusting in underground coal mines. On March 23, 2010, MSHA commemorated the fortieth anniversary of the law by celebrating the decrease in fatalities, injuries, and occupational illness. About two weeks later, the lives of twenty-nine miners were taken at the UBB.

The Mine Safety and Health Act of 1977 added provisions to the Coal Mine Act and expanded the scope of coverage to include mines other than coal mining operations. Once again, legislative changes came after the loss of life: in 1976, two violent blasts at Kentucky's Scotia mine killed fifteen coal miners and eleven rescue workers.[17] Under the law, miners could request an inspection and seek workplace protections. Four inspections were required each year for underground mines. The new act created a new enforcement action to allow MSHA to shut down areas of a mine where inspectors found an "unwarrantable failure" by the mine operator, and it established special requirements for operators found to have a "pattern of violations."[18] MSHA was given the power to request an injunction to close a mine that had engaged in a pattern of violations that constituted a continued hazard to the health or safety of miners. As described later in the chapter, this authority was not used until November 2010, when it was imposed on a mine operated by Massey in Kentucky.

In 2006, after twelve miners were killed in a methane explosion at the Sago mine in West Virginia, the Mine Improvement and New Emergency Response (MINER) Act was passed; the new law strengthened provisions to improve miner safety for the first time in nearly thirty years. It required operators of underground coal mines to improve their accident preparedness plans and develop emergency response plans for regular review and updating. Also required were refuge chambers within one thousand feet of the nearest working face of the mine and tracking and communication devices for miners. The MINER Act set a deadline of June 2009 for underground coal operators to have functioning wireless communication and tracking systems, but fewer than half of the active mines had them at the time of the UBB explosion in March 2010; only about 20 percent of the installation was finished at UBB before the disaster. The law required mine operators to notify MSHA of all incidents that posed a reasonable risk of death, and it raised the amounts for criminal fines and civil penalties.

The laws worked to reduce fatalities in coal mines. In 1977, 273 mining fatalities were recorded in mines across the country; in 2013, MSHA reported that a total of 42 miners died, 20 of them in coal mines.[19]

In sum, coal mining operations remain a major component of US electricity generation. However, coal producers are now challenged by renewable and natural gas options that are more economically viable. The Appalachian region is especially vulnerable to economic downturns, as it does not have the advantages of the extensive and more easily attainable coal reserves in the Western region. Thus, coal mining in Appalachia is threatened, not by regulations aimed at protecting miners and the environment, but by the economies of scale of Western coal, the closing of antiquated coal-fired power plants, and the price competitiveness of natural gas. Perhaps nowhere is this more evident than in the Coal River Valley in West Virginia. Here the battle for cheap coal production has played out in the "hollers" of the mountainous region of the state, and it is the place where the story of the Upper Big Branch mine explosion begins—at the massive operation that was Massey Energy.

SETTING THE STORY: LOCATION AND OPERATIONS AT THE UPPER BIG BRANCH MINE

The Upper Big Branch mine was nestled in West Virginia's Coal River Valley, just forty-six miles from the capital city of Charleston, in Raleigh County. Performance Coal Company, a subsidiary of Massey Energy, operated the mine. At the time of the explosion, Massey Energy was the fourth-leading coal producer in the country, and the largest in the Appalachia region. Massey had more than forty subsidiaries, all of them overseen by CEO Don Blankenship.

Massey operated both surface and underground mines in West Virginia. For example, one of Massey Energy's huge mountaintop removal operations, the Edwight mine, was just a few miles down the road from the Upper Big Branch mine. Environmentalists had focused their attention on Edwight, seeing this mine as a serious threat to the health, safety, and environment of residents of the area.[20] A pond storing 3.8 billion gallons of toxic sludge from mine tailings in an earthen dam at Edwight threatened to break and poison the water of several Appalachian communities. In March 2010, local residents petitioned the West Virginia Department of Environmental Quality (WVDEQ) to hold a public hearing on closing Edwight. The WVDEQ admitted that the mine had been cited nearly three dozen times for violations of environmental regulations, but refused the citizens' request. The Edwight mine captured most of the media attention as well, until the following month, when, on April 5, an enormously powerful blast rocketed through nearly three miles of underground tunnels deep inside another Massey Energy mine in the Coal River Valley.

The UBB mine had avoided most media and citizen attention in part because it was an underground mine, with fewer noticeable surface impacts. The mine was important to Massey, and Massey was eager to see it run coal. The UBB tapped the Eagle Seam of metallurgical coal, a type of coal used for iron and steelmaking. In 2010, metallurgical coal was in high demand as an export to countries with rapidly growing economies, such as China and India. The Eagle Seam was prized as well for its ability to fire steam generators at electric power plants.

Massey began underground coal mining at the site in 1994, when the West Virginia Office of Miners' Health Safety and Training (WVMHST) issued it a permit. The mine represented a good investment for Massey: in its first year, it produced 13,000 tons of coal as a room-and-pillar mine. However, production dramatically improved with the introduction of longwall mining techniques in 1997. When the longwall went into service, coal production at the mine increased to just over 3 million tons annually. When longwall mining ceased for a time at the mine in 2006, owing to a lack of developed panels to mine, the longwall equipment was shifted to Castle mine, another Massey mine. Eventually, bad mining conditions at the Castle mine made it difficult to successfully mine using the longwall machine, and it was returned to the UBB in September 2009. With the return of the longwall equipment, production increased again. From 1994 to 2010, the mine produced approximately 41.4 million tons of coal.[21]

The longwall at the UBB mine normally produced coal seven days a week, with three crews rotating in the production schedule. The longwall face setup consisted of a massive, 90-ton Joy & LS shearer, shields, and a loader unit. It was capable of producing 2,300 tons of coal an hour.[22] At the time of the explosion, mining had progressed more than a mile, about 5,400 feet horizontally, at a depth of over 1,000 feet.

Moving the longwall from the Castle mine proved problematic, as the new setup for longwall mining at the UBB had unresolved ventilation and drainage issues.[23] The importance of proper ventilation cannot be overemphasized. Ventilation not only supplies breathable air but also vents dangerous gases, notably methane, as coal is excavated. When mixed with air, methane can be highly explosive, even in small concentrations. Federal regulations limit levels of methane to just 1 percent in the underground air. Coal dust is also a danger, as it can act as a fuel source. If the volatile combination of coal dust and methane ignites, deadly carbon monoxide is produced.[24] Thus, underground mines require huge fans to circulate the air and provide fresh air to miners. UBB was ventilated with three fans—two to blow air into the mine, and one to exhaust air.

The federal Mine Safety and Health Act requires ventilation based on the amount of methane "liberated" or released into the mine. The UBB mine was in the highest-risk category under the law because it liberated over one million cubic feet of methane a day. Outbursts of gas had occurred in the mine near the longwall at least three times prior to April 5 explosion: in 1997, 2003, and 2004.[25] The 1997 incident caused a series of small explosions and forced an evacuation of the mine. Other protections include self-contained self-rescue devices (SCSRs), which are individual oxygen masks, and rooms in the tunnels stocked with additional breathing equipment. These measures, however, cannot replace adequate ventilation in an underground mine.

THE EXPLOSION: A MINER'S PREMONITION

The day before the mine explosion was Easter Sunday, a rare day when the mine was largely idle as workers relaxed and spent time in paschal activities. However, Gary Wayne Quarles, a thirty-three-year-old tail side shearer operator on the longwall of the UBB mine, did not enjoy the day. Instead, he worried about the mine. He told his close friend Michael Ferrell that something bad was going to happen at the mine, a sentiment he shared with two other friends the previous day. In testimony before the Governor's Independent Investigation Panel (GIIP), Ferrell recalled that Quarles, lamenting the lack of air in the mine, had said: "We ain't got no air. You can't see nothing. Every day, I just thank God when I get out of that coal mine that I ain't got to be there no more. . . . I'm just scared to death to go to work."[26]

Sadly, Quarles did not act on his premonition. On Monday morning, he and other day-shift crews of miners arrived as scheduled. It began as a normal day. By 7:00 a.m., at least forty-five workers were underground. Supervisors had called in the pre-shift reports—none of them recorded any major problems. But then again, foremen knew that it was important to run coal at Massey, so they often did not fully inspect operations, as required by MSHA regulations. For example, a foreman at the longwall, Jeremy Burghduff, did not conduct the required pre-shift examination and left the pump crew without a multi-gas detector. Extensive investigations after the explosion revealed that this was not unusual: Burghduff often failed to check for hazardous conditions and usually left his monitor, which might have alerted him to concentrations of dust or gas, turned off.[27]

Strangely enough, both state and federal mining inspectors also arrived the morning of April 5 to conduct quarterly inspections. State inspector Wayne Wingrove wrote three violations that day, one of which was for low air; MSHA inspector John Syner, after a pre-inspection meeting with mine managers and a brief safety talk to employees, issued two violations—neither one related to the amount of coal dust, the adequacy of ventilation, or the presence of methane.[28] Neither inspected the area near the explosion; both inspectors left before 2:00 p.m. that day.

The miners themselves, however, echoing Quarles's concerns, observed that there was no air deep underground. Several noticed that what little air there was in the mine was headed the wrong direction and carrying coal dust deeper inside the mine. The water pumps at the mine were not working properly, and men were knee-deep and higher in water. Other problems began at 10:00 a.m., when the hinge pin on the ranging arm of the longwall machine came loose and the machine could not operate. It took some time for repairs, but at 2:42 p.m., longwall head-gate operator Rex Mullins reported that they were again running coal. If only the machine had stayed broken, those miners might have lived to see another day.

The shift change had begun, and crews were headed out of the mine. One longwall crew team began the long trip to the surface, unaware that the crew remaining below had cut power to the longwall at 2:59 p.m., most likely in an attempt to stop an explosion.[29] Though investigators never would know for sure, they speculated that the men at the longwall witnessed something ominous and tried to avert disaster by turning off the machine. Less than ninety seconds later, a massive explosion rocketed through the mine, rolling through underground tunnels with such force that some miners were decapitated or thrown to the top of shaft ceilings. One of the survivors, Steven Smith, would later liken the explosion to being in the middle of a tornado, and witnesses would say that it sounded like thunder. White smoke started pouring out of the portals.

MSHA officials and State investigators would later conclude that the ignition point for the blast was the tail of the longwall. As the shearer hit sandstone, it sparked. The spark ignited a pocket of methane gas, creating a fireball. In turn, the fireball spread to the methane that lingered in the "gob" (the area where the coal had been extracted). The fireball traveled into the tailgate area, where it combined with coal dust that provided the fuel source for a second, more forceful and deadly explosion that blasted through more than two miles of mine tunnels and shafts.

Quarles's premonition had proved all too accurate. He and the rest of the longwall day-shift crew were killed. Others tried to flee, but could not outrun the blast and its fierce power. Miners not directly hit by the force of the explosion were killed by high concentrations of carbon monoxide gas. The gas hit them fast; few had time to react or to put on their SCSRs. In total, twenty-nine out of the thirty-one miners were killed. Mine rescue teams would later report that the protective refuge chambers near the longwall had not been deployed—the explosion happened so quickly that no one had time to reach the ventilated rooms containing survival supplies.

Emergency crews gathered immediately, in full rescue mode, hoping for survivors. The initial report was that twenty-five miners had been killed and four men were missing. Hopes for the missing men being found alive dimmed the following day when rescue teams reported that their gas detectors were over-the-range for carbon monoxide and methane. They were forced to abandon their search for survivors. Shafts were drilled to allow the dangerous gases to dissipate, but rescue efforts were hampered for days, as gas readings at the borehole continued to show explosive and lethal levels of gas. Late on April 9, a final briefing for families was held at the mine site, where Governor Joe Manchin and Massey officials informed them that the bodies of the four men had been found. It would take until April 13 for all of the bodies to be removed from the mine, and it would be more than two

Photo 4.1: A makeshift memorial was set up outside the public library in Whitesville, West Virginia, to remember the twenty-nine miners who died in the explosion at Massey Energy's Upper Big Branch mine. *AP Photo/Amy Sancetta.*

months before MSHA investigators could safely enter the mine. Quarles's body was found along with three others about a third of the way down the longwall head gate.

THE CAUSES AND CHARACTERISTICS OF THE UBB EXPLOSION

The tragedy of the UBB explosion shares many of the characteristics of industrial disasters already discussed in the stories of the BP oil spill and the Bhopal "Night of the Gas." Like both of those disasters, the UBB accident was not a "normal" one. Though deep underground mining is a dangerous activity and accidents are expected, the culture at Massey, largely nurtured and promoted by its CEO, Don Blankenship, was responsible for the tragedy.

A History of Disregard for Safety and "Normal" Accidents

Shortly after it finished its investigation of the "Industrial Homicide" at the UBB mine, as it titled its report, the United Mine Workers of America

(UMWA) observed, "There were many factors that led to this disaster. But there is only one source for all of them: a rogue corporation, acting without real regard for mine safety and health law and regulations that established a physical working environment that can only be described as a bomb waiting to go off."[30]

The Governor's Independent Investigation Panel (GIIP) drew similar conclusions about the culture of production over safety at Massey Energy. The panel drew on Diane Vaughan's organizational theory of the "normalization of deviance" (described in Chapter 1): the gradual process by which unacceptable practices become the norm in an organization, creating a culture that permits, even encourages, behaviors that deviate from standard safety protocols. In describing a litany of unsafe practices, the panel's report struck at Massey's "production at all costs" culture: "Most objective observers would find it unacceptable for workers to slog through neck-deep water or be subjected to constant tinkering with the ventilation system—their very lifeline in an underground mine. Practices such as these can only exist in a workplace where the deviant has become normal, and evidence suggests that a great number of deviant practices became normalized at the UBB mine."[31]

The GIIP also refuted Massey's claim that this accident was simply due to an unfortunate and unexpected release of methane gas, stating: "This history of inadequate commitment to safety coupled with a window dressing safety program and a practice of spinning information to Massey's advantage works against the public statement put forth by the company that the April 5, 2010, explosion was a tragedy that could not have been anticipated or prevented."[32]

Evidence suggests that this disregard for safety and the culture it created emanated from the very top of this corporation, with CEO Don Blankenship. Blankenship oversaw all aspects of this organization and micromanaged the decisions of managers, even signing off on the hiring of every worker and on small purchases.[33] Despite the vast holdings of Massey Energy, Blankenship was a hands-on executive who flew by helicopter to check on production numbers at every mine. Mine supervisors were required to send production reports every thirty minutes, with explanations if numbers were low. Not surprisingly, Blankenship was no fan of either the United Mine Workers of America or the MSHA. He called environmentalists "greeniacs" and considered climate change a hoax. But his union-busting tactics and financial savvy found favor with the company, and he had become chairman and CEO of Massey in 1992—just ten years after joining the company.

The "Dark Lord of Coal Country" and Massey's Bottom Line

On the one hand, Blankenship's handling of Massey Energy as CEO was impressive. He transformed Massey from a "sleepy old coal company" into the most powerful economic and political machine in Appalachia, with more than fifty-six mines employing nearly six thousand workers mining forty million tons of coal each year.[34] On the other hand, dozens of accounts, including the testimony of people inside of Massey, identified him as calculating, with a disregard for any worker safety or environmental standards that might negatively affect mine profitability. An article in *Rolling Stone* christened Blankenship "the Dark Lord of Coal Country," noting that he embodied

> everything that's wrong with the business and politics of energy in America today—a man who pursues naked self-interest and calls it patriotism, who buys judges like cheap hookers, treats workers like dogs, blasts mountains to get at a few inches of coal and uses his money and influence to ensure that America remains enslaved to the 19th-century idea that burning coal equals progress. And for this, he earns $18 million a year—making him the highest-paid CEO in the coal industry—and flies off to vacations on the French Riviera.[35]

Blankenship pursued increased corporate profits not only by shaving safety protocols but also by expanding Massey's reach into Appalachian coal. His expansion strategy for Massey included buying smaller coal companies. If these companies resisted Massey's takeover bid, Blankenship would engage in maneuvers that forced the targeted company into bankruptcy. One infamous example was Harman Mining. Harman's president, Hugh Caperton, sued Massey, arguing that Blankenship had set out to destroy the business by purchasing coal reserves around his mine and breaking the contract with the company that produced Harman coal. In 2002, a jury found that Massey had committed fraud in its business dealings and awarded $50 million in punitive damages to Caperton.[36]

Undaunted, Massey appealed the decision to the West Virginia Supreme Court of Appeals. To help ensure that the appeal would go his way, Blankenship spent $3 million in a smear campaign to oust Justice Warren McGraw and elect a conservative justice, Brent D. Benjamin. The $3 million campaign financed by Blankenship was more than three times as much as Benjamin's own campaign contributed.[37] Benjamin won his election just in

time to hear the appeal. Refusing to recuse himself, Benjamin joined the
3–2 majority that overturned the $50 million verdict against Massey. One of
Benjamin's colleagues on the court, Justice Larry V. Starcher, commented,
"We have one justice who was bought by Don Blankenship. It makes me
want to puke."[38] Blankenship discounted these criticisms, noting that he had
spent $3 million from his own pocket for television ads aimed at defeating
the incumbent, not in favor of electing Benjamin. "Eliminating a bad politician makes sense," he quipped in an interview with the *New York Times*.
"Electing somebody hoping he's going to be in your favor doesn't make any
sense at all."[39]

On appeal, the US Supreme Court held that due process required that
Justice Benjamin recuse himself and sent the case back to the West Virginia Supreme Court, citing an "extreme" conflict of interest. The case was
reheard by the West Virginia Supreme Court, but overturned on jurisdictional issues; Caperton was told that he should pursue his claims in Virginia. Blankenship's electioneering tactics had prevailed, but not without
drawing extensive media attention, court scrutiny, and the attention of the
novelist John Grisham, who used the case as the basis for his book *The Appeal*. When the case was litigated in Virginia in 2014, a jury awarded Harman Mining $4 million in damages, finding that Massey had driven the
company out of business through its elaborate scheme.[40] The amount was
far less than the $90 million sought by Harman, but stood as an example of
how far Massey and its CEO would go to get their way.

In another example, Massey Energy supervisors pled guilty in 2002
to failing to perform pre-shift examinations at the White Buck Number 1
mine, as required by federal regulation. But this was to be expected, given
the "production at all costs" culture at Massey. Failure to examine mine
conditions prior to a new shift of workers arriving became commonplace
in Massey mines. After all, any problems identified would need reporting,
maybe even fixing, and might stop coal production entirely.

Perhaps nothing illustrates Blankenship's singular focus on the bottom
line better than the infamous "run coal memo" to deep-mine superintendents on October 19, 2005. The memo was startlingly clear in its direction:
"If any of you have been asked by your group presidents, your supervisors,
engineers or anyone else to do anything other than run coal (i.e.—build
overcasts, do construction jobs, or whatever) you need to ignore them and
run coal. This memo is necessary only because we seem not to understand
that coal pays the bills."[41]

Just three months after that memo was sent to Massey mines, in January 2006, a deep fire caused by an improperly maintained conveyor belt at

Aracoma Coal, a Massey subsidiary, took the lives of two miners. The widows took Massey to court, arguing that Blankenship and other executives knew about the conveyor belt and negligently allowed mining to continue. The "run coal" memo did not sit well with West Virginian mining communities once it was made public, and to avoid a court battle, Aracoma Coal settled the wrongful death suit in 2008. Aracoma Coal eventually pled guilty, in 2009, to ten criminal charges and paid $4.2 million in criminal fines and civil penalties. As part of the deal, prosecutors agreed not to pursue charges against company executives, prompting the plaintiffs' attorney, Bruce Stanley, to later lament: "Sadly, aggressive prosecution against upper management in the Aracoma case might have spared us the horror of UBB."[42]

Massey was not done with lawsuits stemming from Aracoma. A shareholder lawsuit claimed that Blankenship and the Massey Energy board of directors were devaluing the stock price by "failure, among other things, to implement adequate internal controls to ensure the company's compliance with applicable laws and regulations concerning worker safety and environmental protection."[43] Massey settled once again, and the court order required that the company create a safety, environmental, and public policy committee and the positions of vice president for best environmental practices and vice president for best safety practices. These officers and the committee were tasked with monitoring the company's safety and environmental practices and creating a process by which all employees, suppliers, customers, and advisers could provide confidential information regarding unsafe, illegal, or unethical conduct by the company in complying with safety and environmental regulations.

The court order to create a high-level safety and environmental committee apparently did little to stop the destructive environmental and worker safety practices at the Upper Big Branch mine. Testimony after the UBB explosion revealed a long list of purposeful attempts by Massey to evade regulatory requirements. Among the saddest in terms of health effects for the miners was the falsification of the dust sampling data used to enforce federal black lung protections for miners. One way to understate exposure was to have miners wearing the sampling devices called dust pumps sit in the fresh air intake tunnels. Miner Bruce Vickers told investigators that when he wore the pump, his managers would "keep me in the intake, in fresh air," and that he was told not to go into the dusty parts of the mine. Others had similar accounts. For instance, Mike Kimblinger, a construction foreman, recounted in sworn testimony that he was "told to stay away from the dust and not do certain things" while wearing the dust pump. Mark Edwards, who ran a shuttle car at UBB, told investigators that the company would shut down one of the two continuous miners (the longwall machines) during the sampling. "They didn't care about coal that day. Any other day, we're running two miners [machines]. It was so dusty down there it was awful."[44]

Autopsies of the twenty-four victims of the explosion who had sufficient lung tissue to sample revealed that 71 percent of them had black lung disease. Black lung, or coal worker's pneumoconiosis, is an irreversible and potentially deadly disease caused by exposure to coal dust. This compares to a national black lung rate of 3.2 percent among active underground miners and a rate in West Virginia of 7.6 percent.[45] In short, miners at the UBB mine were ten times more likely to contract this often fatal disease than their counterparts in other mines. The UBB explosion may have killed twenty-nine miners, but anyone working in high levels of coal dust also confronted the very real possibility of dying from black lung.

Another kind of dusting, rock dusting with crushed limestone, was not a priority at UBB. Rock dusting is one of the most basic elements of safe mining practices: rock dust dilutes the explosive nature of coal dust, thus preventing coal dust from turning a flare-up into a major explosion. Investigations revealed that Massey provided only a two-man crew to dust the massive underground mine, and then only on a part-time basis. This essential safety chore was difficult at UBB to begin with, owing to the size of the mine and inadequate staffing, but it was further complicated by antiquated and ill-functioning equipment. The GIIP investigation suggests that the rock-dusting machine was acquired in the 1994 purchase of the mine and had not been maintained. With a cantankerous old piece of equipment, UBB miners testified that they spread rock dust by hand on the floors and walls of the working sections of the mine; they often left the roof to accumulate coal dust, since they were unable to reach it.

Other issues went unresolved. Records revealed that, in addition to the dearth of efforts to control high levels of coal dust, nothing was done to fix ventilation problems. Safety mechanisms, including methane detectors, were disabled, so that production could continue without taking time to make repairs. Workers who questioned safety conditions were intimidated. Tailgate 22 foreman Brian Collins testified that when he delayed coal production for an hour until the mine could be adequately ventilated, he was suspended for three days for "poor performance."[46] Collins was not alone. "No one felt they could go to management and express their fears," a miner named Stanley Stewart testified after the disaster. "We knew that we'd be marked men and the management would look for ways to fire us. Maybe not that day, or that week, but somewhere down the line, we'd disappear. We'd seen it happen. I told my wife I felt like I was working for the Gestapo at times."[47]

The legacy of poor practices had resulted in MSHA citations far exceeding the average for similar mines. The UBB had been cited for 458 safety violations in 2009. Of those, 10 percent were citations for "unwarrantable failure to comply," which were reserved for cases of gross negligence. This rate was five times the national average.[48]

Inadequate Planning and Preparation

Massey not only ignored or minimized safety and environmental protocols but also failed to plan for mine operations. Sometimes the company failed to settle on a plan at all. For example, in the seven months leading up to the disaster, UBB management submitted to MSHA more than forty revisions to the mine's ventilation plan for approval. This haphazard approach meant that the ventilation system—the lifeline for workers underground—was not systematically engineered. It also made it difficult for mine operators to stay on top of changes. Had the emphasis been put on properly planning for ventilation of the mine instead of running coal, the fateful events of April 5 might not have occurred at all.

Massey also failed to monitor equipment, such as the rock-dusting machine. A mine the size of the UBB should have included a plan for drilling a borehole to allow for the speedy and regular delivery of rock dust. No such borehole or plan for one existed at the UBB.[49] Without a borehole, the dusting crew had to make the laborious one-hour trek to the outside of the mine to refill the duster. Even more troubling, Massey had no plan for adequately staffing the all-important rock-dusting operation (using just a two-man crew) and, as mentioned earlier, did not perform regular maintenance on the duster, a situation that was immediately evident to MSHA inspectors after the accident who tried to start it up. The lack of attention to rock-dusting and the absence of a plan to restore old equipment made the UBB a very risky mine.

Keeping vital equipment in running order is important. So, too, is having emergency evacuation procedures in place. In March 2010, after federal inspectors had cited Massey for not providing adequate and clearly marked escape routes in the event of emergencies, the company was fined for failing to have adequate escape route plans.[50]

AN INEFFECTIVE REGULATORY PRESENCE AND THE MSHA

In this story, it could be argued that inspectors were doing their job. Far from the cozy relationships the MMS had with offshore drillers, state and federal inspectors were not reluctant to engage Massey over safety and environmental violations. For example, concerned about inadequate rock dusting, inspectors cited UBB for this violation fourteen out of the fifteen months preceding the explosion, with nearly half of the citations noting significant and substantial violations for coal dust accumulation.[51] This was nothing new to the miners, who testified that few areas of the mining operation were adequately protected from a coal dust explosion. UBB fire bosses and foremen would note the need for rock dusting on their pre-shift examinations, but adequate rock dust was not applied. In the three weeks before the disaster, in follow-up to the 561 "needs dusting" notations in pre-shift reports, only 11 percent of the rock dustings requested were completed.[52]

MSHA issued safety violations at the mine at twice the national average.[53] In the year before the tragedy, parts of the mine had been closed more times than any other mine in the country for safety violations and cited more than forty-eight times for air-related problems.[54] Adequate ventilation in any underground mine is crucial, but the UBB was in a geological formation known to contain lots of methane. Moreover, the mine shafts stretched for miles underground, leaving miners especially vulnerable in the event of an explosion. Most telling, MSHA issued sixty-one withdrawal orders in the sixteen months before the explosion, temporarily shutting down parts of the UBB. Sadly, even though this number of withdrawal orders was unheard of in the mining industry, MSHA did not completely close the mine.[55]

Massey did little in response. While the company had been fined over $43 million for safety violations between 2005 and 2010, it had paid just over $10 million, all the while conducting business as usual. The company contested 65 percent of its violations between 2006 and 2009.[56] The practice of challenging fines and avoiding punishment led to a significant backlog of legal actions, thus allowing violations to continue.

On the other hand, it could also be said that inspectors were not effectively doing their job. Despite the UBB being cited hundreds of times in the years before the explosion, MSHA inspectors never issued a "flagrant" violation against UBB. This tough enforcement tool provided to the agency by Congress as part of the MINER Act of 2006 increased the fines that could be levied against companies that "repeatedly fail to make reasonable efforts to eliminate a known violation of a mandatory health or safety standard . . . reasonably expected to cause death or serious bodily injury."[57] MSHA also failed to use its power to put Massey in a "pattern of violation" category for its history of substantial and significant violations. A mine so categorized can result in miners being ordered out of the mine if any subsequent violations are found. Instead, MSHA notified Massey that there was a "potential" pattern of violation, which allowed the company to avoid stiffer sanctions.

Perhaps most disturbing is that MSHA officials were aware of widespread lapses of enforcement. An internal audit of twenty-five field offices the year before the UBB explosion found a number of problems, including incomplete inspections, failure to monitor mines that were liberating high amounts of methane, and inadequate reviews to see if the mines had taken appropriate corrective action.[58]

This conclusion was also reached by the GIIP team. In closely examining the conduct of MSHA, the governor's panel identified four failures of MSHA regulatory oversight. The first was disregarding the risk of methane outbursts at the UBB. The mine was "gassy," liberating one million cubic feet of methane every twenty-four hours. Moreover, the mine had experienced three major methane events; Massey officials considered them anomalies, but MSHA inspectors should have recognized them as the result of serious conditions warranting special precautions. The Mount Hope field office of the MSHA had taken no action.

Second, Mount Hope field office inspectors did not act to shutter the mine even when confronted with the precarious state of the ventilation system. During the GIIP investigation, the assistant director of the Mount Hope field office noted that the president of Performance Coal Company (the Massey subsidiary operating the mine) would routinely ask that ventilation plans be approved quickly. Investigators suggested that instead of giving the UBB special consideration, the field office should have conducted an in-depth review.

Third, the report noted that MSHA could have leveraged its regulatory authority to force Massey to improve its technology. One example that might have prevented the UBB explosion would have been requiring the use of a meter to calculate the explosive potential of the coal dust. In addition, the intermittently operable rock duster could have been replaced under the regulatory authority given to the MSHA under the Coal Mine Act.

Finally, the most damning observation made by the governor's panel was that the US mine safety system had been allowed to atrophy. The panel observed that the ultimate failure of MSHA was its inability to see the big picture or to link the many potentially catastrophic failures taking place at the mine. It admonished the agency: "The ability to stand back and take a long look—to see the red flags, to connect the dots—and the ability and willingness to take quick action when necessary distinguishes a regulatory agency which can prevent disaster from one which only reacts."[59]

In sum, the four characteristics of industrial disasters were present in this story. The Upper Big Branch explosion was but the most tragic in a long history of regulatory violations at Massey under Blankenship's leadership. In the hazardous business of mining, Massey would come to be known as the most dangerous mining company in the nation. In the decade before the explosion, no US coal company had a worse fatality record than Massey.[60] Between 2000 and 2009, the government cited Massey for 62,923 violations and proposed nearly $50 million in fines. Blankenship dismissed these violations as a normal part of the mining process, especially considering the difficulty of underground mining, but US Department of Labor solicitor M. Patricia Smith disagreed: "Everybody gets violations sometimes, but when you actually look at comparing apples to apples and oranges to oranges, serious citations to serious citations and similar mines, Massey had a very bad record."[61]

Massey also led efforts to expand mountaintop removal mining. Seeing competitive pressures from the big mines in the West, Blankenship and other Appalachian coal executives pushed for mountaintop removal because blasting was cheaper than digging, but the environmental consequences were serious. Interviewed for the *Rolling Stones* investigation, UMWA president Cecil Roberts lamented, "Blankenship has probably caused more suffering than any other human being in Appalachia."[62]

AFTER THE EXPLOSION: THE END OF A COMPANY

The worst mine disaster in over forty years captures media attention, the attention of politicians, and the attention of regulators. This disaster was too big, attracting too much attention, for Don Blankenship to continue to work his magic in West Virginia. Nevertheless, even under fire from regulators and shareholders as Massey's stock price plummeted, the board of directors quickly circled the wagons, retaining a public relations firm two weeks after the explosion and issuing a statement that "Mr. Blankenship has the full support and confidence of its members."[63]

The public relations campaign did little, however, to stop the furor over the UBB explosion, and much of it was directed at Blankenship. Under Blankenship, the company's annual revenues had doubled, but those profits had come with significant environmental and human costs. Even before the human tragedy at UBB, Massey had amassed thousands of state and federal safety violations under Blankenship's reign—many of which the company was still fighting at the time of the explosion.[64] Ongoing federal and state investigations at the UBB mine would bring an unprecedented series of new fines and regulatory violations, not to mention litigation from the families of the victims. In 2010, Massey announced a net loss of $166.6 million, compared to a profit of over $100 million in 2009.

Unprofitable and mired in controversy as Massey was, its undervalued stock price and control of vast Appalachian coal reserves—including the largest reserves of metallurgical coal in the country—made the company an attractive takeover target.[65] Blankenship wanted to fight any takeover attempts as tenaciously as he had fought the unions and regulators, but this was the final straw for shareholder groups, who pressed for Blankenship's departure. Like BP's CEO Tony Hayward, Blankenship was no longer an asset to the company. Though the unexpected departure was described by Blankenship and the board as a resignation, it seems clear that Blankenship was encouraged to go.

Massey had dealt itself its own death blow. Alpha Natural Resources took over in a merger on June 2, 2011, and Massey ceased to exist as a company. The takeover made Alpha one of the three largest coal companies in the country: worth $15 billion, it now operated more than 110 mines and held five billion tons of coal reserves.[66] If anyone felt any sympathy for Blankenship or the Massey board of directors, it probably evaporated once the terms of the merger with Alpha were disclosed. Eighteen current and former Massey executives and board members shared $196 million in salary, benefits, severance, pension, retirement, and deferred compensation. Blankenship reportedly received the lion's share—more than $86 million.[67]

Just a few days after the takeover by Alpha, Massey issued its own investigative report. The company found that the explosion was the result of a massive inundation of natural gas, not the buildup of coal dust or poor safety practices at the mine. The report stated: "The government has ignored compelling evidence of a natural disaster and, instead, focused single-mindedly on any factors that were conceivably within Performance's control. Consequently, MSHA has disregarded all scientific data demonstrating that a massive gas inundation caused the explosion, preferring instead to point to coal dust, which the government typically believes to be within an operator's scope of responsibility."[68]

Massey experts claimed that ventilation changes ordered by MSHA were also to blame, and the company went so far as to accuse MSHA of manipulating testimony in its investigation, even of coercing mine staff to destroy evidence. The report concluded that "the government cannot currently say with any reasonable confidence that Performance management or its members caused the UBB tragedy."[69] This effort seemed as bizarre as it was pointless, especially given that three separate independent investigations pointed to far different conclusions. Alpha, in negotiations with the Justice Department over the violations at the UBB, was not inclined to support the findings of Massey's internal report.

The explosion changed many lives, but the families of the victims were most deeply affected. The company's first offer to each of the families of the twenty-nine victims was $3 million. Gary Quarles, the father of the miner who had the eerie premonition about the UBB the day before the explosion, called the offer of compensation for the death of his only son a "slap in the face." A miner working for Massey himself, Quarles knew the corporate culture well. "'Production first, safety last, haul the coal or haul your a——,'" Quarles said, reciting what he and other miners believed was the true Massey creed. "You were just a number to them. If you were producing, fine. If not, you were just dirt under their feet."[70] Seven families took Massey's offer, described in a *Washington Post* article as "meaningless millions," while others filed suit against the company.[71]

GOVERNMENT INVESTIGATIONS AND CRIMINAL PROSECUTIONS

Three government investigations were conducted in the months following the UBB tragedy. The first to issue its report was the Governor's Independent Investigation Panel, whose findings permeate this account of the explosion. The GIIP report began with a simple statement: "The explosion at the Upper Big Branch mine could have been prevented."[72] It concluded with the same sentiment: the explosion was "a completely predictable result for a company that ignored basic safety standards and put too much faith in its own mythology."[73] The GIIP investigators identified failures of basic safety systems—including poor ventilation, meager rock dusting, and improperly maintained water sprays on equipment—that could have prevented the initial ignition of methane gas. The report directed most of its ire at Massey, labeling the "Massey Way" as operating a company "well known for causing incalculable damage to mountains, streams and air in the coalfields; creating health risks for coalfield residents by polluting streams; injecting slurry into the ground and failing to control coal waste dams; . . . using vast amounts of money to influence the political system; and battling government regulation regarding safety in the coal mines and environmental safeguards for communities."[74]

MSHA issued its own eight-hundred-page report on December 6, 2011, which it described as the most extensive investigation of a mining disaster in modern times.[75] The report found "multiple examples of systemic,

FIGURE 4.1: TIMELINE OF THE UPPER BIG BRANCH MINE EXPLOSION

1992: Don Blankenship is appointed chairman and CEO of Massey.

1997: Blankenship tries to pressure a competitor, Harman Mining, into bankruptcy by buying the coal processing company. The CEO, Hugh Caperton, will sue Massey and Blankenship in 1998.

2000: A coal slurry spill in Martin County, Kentucky, releases three hundred million gallons of toxic sludge into nearby streams.

2002: A jury awards Caperton $50 million in damages. Massey appeals.

2006: A fire at the Aracoma mine (owned by a Massey subsidiary) kills two men. The Mine Safety Health Administration (MSHA) investigation shows that the conveyor belt was improperly maintained. Three months earlier, Blankenship wrote a memo to all mine superintendents demanding that they "run coal" instead of dealing with safety issues.

2007: The West Virginia Supreme Court overturns the verdict for Caperton. A year later, photos showing Blankenship and the chief justice of the West Virginia Supreme Court vacationing in the Riviera prompt a rehearing.

2008: Aracoma Coal pleads guilty to ten criminal charges and pays $4.2 million in fines. The Massey operation agrees to resolve 1,300 safety violations. The judge for the case questions whether the penalty is sufficient.

2009: The US Supreme Court rules in *Caperton v. A. T. Massey Coal Co.* that the state judge whose election was backed by Blankenship should have recused himself. The Mine Safety Health Administration (MSHA) cites five hundred violations at the Upper Big Branch mine, but does not close the mine. Mine records kept by Massey workers but not shared with inspectors reveal persistent and explosive levels of methane, inadequate ventilation, and improper treatment of coal dust.

April 5, 2010: The Upper Big Branch Mine explodes, killing twenty-nine miners, the worst coal mine disaster in forty years. West Virginia governor Earl Tomblin appoints the Governor's Independent Investigation Panel (GIIP) to determine the cause of the explosion.

May 2010: Federal prosecutors announce a criminal investigation.

December 2010: Blankenship retires from Massey amid public pressure, but with a retirement package worth over $80 million.

2011: Massey Energy agrees to be taken over by Alpha Natural Resources, which pays $210 million to the families of the victims and the government for years of Massey violations.
MHSA and GIIP issue their investigative reports on the UBB explosion, revealing multiple examples of intentional efforts by Massey to avoid compliance with safety and health regulations.

2014: A federal grand jury indicts Don Blankenship.

2016: Blankenship is sentenced to one year in prison and fined $250,000 for conspiring to violate federal mine safety standards. MSHA observes the sixth anniversary of the explosion.

2017: Blankenship is released from prison.

intentional and aggressive efforts by Massey to avoid compliance with safety and health standards, and to thwart detection of that non-compliance by federal and state regulators."[76] The stunningly brazen conduct of Massey executives and officials was now on full display. In assessing the root causes of the explosion, the report identified numerous management violations of federal law, including intimidating miners to prevent federal inspectors from receiving evidence about safety and health violations; providing advance notice of inspections to hide violations from federal enforcement personnel; and keeping two sets of mine examination books—the doctored set to give to federal inspectors and the "management eyes only" set that recorded the actual hazards at the mine.

Equally egregious were the findings that Massey managers and executives allowed hazardous levels of loose coal and coal dust to accumulate, while also failing to adequately apply rock dust to the mine and failing to comply with approved ventilation plans. Had any one of these issues (rock dusting, ventilation, coal dust accumulation) been adequately addressed, the explosion most likely would not have been so extensive, with such a high cost in human life. This repetitive and willful violation of the law by the operators of the mine, according to the MSHA report, encouraged noncompliance by mine managers. A Massey official suspended a section foreman, according to witness testimony, who delayed production for two hours to address safety concerns. The report noted that if miners initiated production delays to resolve safety issues, they too often faced "threats of retaliation and disciplinary action."[77]

In the aftermath of the UBB explosion, MSHA imposed a record $10.8 million in civil penalties—the largest in MSHA history. It also issued an unprecedented 369 citations and orders on the company, including 21 flagrant violations of safety and health standards. However, none of the fines held the managers or senior executives personally accountable for their actions that resulted in the miners' deaths. Don Blankenship and other members of the board exercised their Fifth Amendment rights and were not interviewed during the investigation.

During its twenty-two-month investigation after the explosion, WVMHST issued 253 violations. Twenty-eight additional violations were issued after the agency audited the company's accident history for failure to notify WVMHST of reportable accidents.[78] Like MSHA, West Virginia investigators found that the explosion was the result of methane accumulating in the gob behind the longwall shields, which probably occurred as the shearer was cutting sandstone roof. Like their federal counterparts, the state agency found that the initial explosion of methane transitioned into a coal dust explosion, which then propagated through an extensive area of the mine. Noting that Massey failed to remove hazards and violations during mine examinations, WVMHST observed that there were indications that no rock dusting had been done at all in certain areas of the mine since the longwall operation began in 2009.[79] The report went on to note that the agency's state statutory language was insufficient to regulate the way coal mines are ventilated, and it called on coal operators to take a more proactive approach to the ventilation of mines under their authority and to not be "so quick to disregard the engineer's professional judgment."[80]

On the same day as the release of the MSHA's final investigative report, the Department of Justice announced a nonprosecution agreement with Alpha Natural Resources.[81] Under the agreement, the government would forgo prosecuting Alpha Natural Resources in exchange for restitution, investments in mine safety, and payment of civil penalties. In return, Alpha agreed to pay $210 million to avoid prosecution, including $46.5 million to the families of the disaster victims and $35 million to resolve Massey safety fines.[82] Alpha agreed to spend $80 million during the next two years on mine safety improvements and to create a $46.5 million mine safety research trust fund. The settlement barred criminal prosecutions against the companies (Alpha Natural Resources, Massey, and Performance Coal), but nothing in the agreement prevented the Department of Justice from bringing criminal charges against individuals responsible for the disaster.

By early 2014, three criminal convictions had been made. Former UBB superintendent Gary May pled guilty in 2013 to a federal conspiracy charge. He was accused of defrauding the government through his actions at the mine, which included disabling a methane gas monitor and falsifying records. After being sentenced to twenty-one months in jail, May asked that the sentence be set aside, arguing that he had been made a scapegoat by Massey's general counsel in order to protect Massey executives.[83] David Hughart, a former Massey executive and division president, pleaded guilty to two federal charges—one felony count of conspiracy to defraud the government and one misdemeanor count of conspiracy to violate MSHA standards. He was sentenced to forty-two months for his role in illegally notifying mine operators about surprise inspections. During his plea hearing, Hughart implicated Blankenship in the conspiracy.[84] Former UBB security director Hughie Elbert Stover was sentenced to thirty-six months after he was convicted in a jury trial of making a false statement and obstructing the government's investigation into the mine disaster.[85]

In February 2014, the federal government determined that Alpha had met or exceeded its obligations under the nonprosecution agreement, but

US Attorney Booth Goodwin's efforts to investigate misconduct by individual officials of the former Massey Energy continued.

Goodwin's efforts would bear fruit. Blankenship was eventually charged on three counts: two counts of making false statements, and one count of conspiring to violate federal mine safety standards. The jury delivered a mixed verdict in 2015, acquitting Blankenship on the more serious felony charges, but finding him guilty of misdemeanor conspiracy. The guilty verdict imposed the maximum $250,000 fine and a one-year jail sentence. Though far less than the thirty-year prison term prosecutors had sought, this sentence is noteworthy because CEOs rarely go to jail. Blankenship served his sentence and was released on May 10, 2017, to serve an additional year of supervised release. The former Massey CEO remained defiant, calling himself an "American political prisoner" and tweeting a challenge to Joe Manchin, now a Democratic West Virginia senator, to "be man enough to face me in public."[86] After retiring from Massey, Blankenship envisioned returning to his status as a coal baron. He incorporated a new venture in Kentucky, the McCoy Coal Group, Inc., but has yet to seek any mining permits for the new company.[87]

POLICY CHANGES AND SHORT-LIVED CONGRESSIONAL ATTENTION

If UBB was a disaster waiting to happen—as indicated by a long history of violations for coal dust, poor ventilation, and other dangerous conditions— why did federal inspectors fail to shut the mine down? "Because nobody shuts one of Don Blankenship's mines down," said a miner in an interview for *Rolling Stone*. "It has never happened. Everyone knows when mine inspectors are coming, you clean things up for a few minutes, make it look good, then you go back to the business of running coal. That's how things work at Massey. When inspectors write a violation, the company lawyers challenge it in court. It's all just a game. Don Blankenship does what he wants."[88]

But not after UBB. As with the BP oil spill, the explosion forced the federal mine safety agency to take a hard look at its efforts. During a hearing before the Senate Health, Education, Labor, and Pensions Committee three weeks after the tragedy, MSHA administrator Joseph Main promised Congress that the agency would be more aggressive in enforcing mine safety laws. MSHA conducted surprise inspections and ordered the evacuation of three other Massey-owned West Virginia mines in the weeks after the UBB tragedy.[89] It took control of the phone lines at two of these mines to ensure that mine foremen were not tipped off about the presence of inspectors, an illegal activity that occurred at UBB. On April 29, just two days after the congressional hearing, two miners were killed when a roof collapsed in a Kentucky mine. Though not owned by Massey, the mine also had a long history of safety problems, with more than 40 closing orders and 840 violations.[90] In November 2010, MSHA for the first time asked a federal judge to shut down Massey Energy's Freedom Mine Number 1 in Kentucky. Though the agency had the authority to seek injunctive relief against mining companies under the 1977 law, it had never used it.[91] Perhaps the specter of additional tragedies at coal mines fortified the agency's backbone, as it did for Congress, which seemed more willing to stiffen rather than relax regulations.

However, while MSHA resolve increased over the years following the disaster, congressional resolve to strengthen national laws waned. For one thing, the BP oil spill dominated media coverage and propelled the related environmental issues onto the national agenda. For another, the coal industry holds enormous political power in state and national government, and its influence would soon reemerge. In the months after the tragedy, it appeared that Congress would pass new legislation closing the loopholes in the current mining laws. Hearings were held on the UBB tragedy in May 2010. The Robert C. Byrd Mine Safety Protection Act (MINER) of 2010 was introduced but failed to pass—as would also happen in 2011 and 2013.

A year after the explosion at the UBB took the lives of twenty-nine miners, J. Davitt McAteer, who had led the governor's independent investigation and also headed the Mine Safety and Health Administration for seven years, was asked if regulations protecting worker safety had changed. He responded that no changes had been made, at either the state or federal level, and speculated that because the BP oil spill had captured congressional attention, there was no impetus to address miner safety. McAteer noted that the MINER Act of 2010 that failed to pass could have dramatically affected mine safety because it put responsibility on mine owners and boards of directors rather than on mine foremen. "As long as we have that disconnect, it's going to give those who want to disregard the law the ability to do so. There needs to be a connection between the wash room at the mine and the board room."[92]

Instead, Representative John Kline (R-MN), the new chairman of the House Committee on Education and the Workforce, chose in 2012 to sharply criticize MSHA and its "broken enforcement regime":

It is difficult, almost impossible, to imagine enforcement personnel missing the inherent dangers of coal dust accumulating throughout the mine. Again, this enforcement error neglected a crucial safety concern that would later enhance the magnitude of this disaster. We have also learned over the last 2 years that other enforcement tools were either poorly used or never implemented. Bipartisan reforms enacted in 2006 created a new category of flagrant violations, yet they were never imposed against Massey.[93]

Absent from this critique, however, was the fact that most Republicans in Congress supported the long-standing commitment to lightening the regulatory burden of companies. The Bush administration was a friend of coal and an advocate for regulatory reforms; this position was supported by many Republicans in Congress and became especially visible after the party gained congressional seats in subsequent elections.

But it was not only Republicans who sought to protect the coal industry. Just six months after UBB exploded, West Virginia governor Joe Manchin—the same governor who had ordered the special investigation into the UBB—directed the state to sue the Obama administration to overturn new federal rules on mountaintop removal mining. Under the new EPA regulations, companies seeking Clean Water Act permits for proposed surface mines would have to demonstrate that their discharge into surrounding waters would not cause significant pollutant increases. To be sure, this time it was the safety of the environment at risk, but the Democratic governor, who was in a special election bid for a seat in the US Senate (which he won), wanted to distance himself from environmentalists and align himself instead with coal mining interests. In announcing his instruction to sue the EPA, Manchin noted that he had been "fighting President Obama's administration's attempts to destroy the coal industry and our way of life in West Virginia." He went on, "We are asking the court to reverse EPA's actions before West Virginia's economy and our mining community face further hardship."[94]

The UBB mine was plugged and sealed permanently on June 20, 2012.[95] A memorial in Whitesville, West Virginia, for the twenty-nine lost miners reads: "Come to me, all you who labor and I will give you rest."[96]

CONCLUSION

Sadly, the Upper Big Branch mine explosion, much like the tragedy in Bhopal and the explosion of the *Deepwater Horizon,* is the story of a villainous company headed by a powerful CEO who wanted to build an empire with little regard for other people or for the environment. Like the Bhopal and

Deepwater Horizon explosions, the UBB tragedy was foreseeable and preventable. All that was required was that Don Blankenship and the managers of Massey mines pay more than lip service to their safety programs. Instead, in the rush to make more money by producing coal at any cost, Massey established a culture that punished anyone who spoke out about hazardous conditions or, worse, who wanted to stop production.

In his interview with *Rolling Stone*, Bruce Stanley, a lawyer who grew up in Mingo County, West Virginia, observed, "One thing that is hard to take about Don Blankenship is how he betrayed his own people. . . . Blankenship could have easily been a hero, not a villain. He could have said to the people of Appalachia, 'Let me show you how to pick yourself up by your bootstraps. Let me show you how to make something of yourself.' Instead he said, 'Fuck it—I'm king.'"[97] As noted in the *Rolling Stone* article, if any of the events at UBB trouble Blankenship, the bottom line provides all the proof he needs of his own virtue. "I don't care what people think," Blankenship once said during a talk to a gathering of Republican Party leaders in West Virginia. "At the end of the day, Don Blankenship is going to die with more money than he needs."[98] That's small comfort to the families whose loved ones died on April 5, 2010.

DISCUSSION QUESTIONS

1. What economic, political, and social conditions facilitated the rise of Don Blankenship, CEO of Massey?
2. What factors account for changes in demand for coal? What effect, if any, will these factors have on coal mining policies?
3. Why do you think Gary Wayne Quarles, the coal miner with a premonition about the unsafe conditions at the UBB mine, still went to work on Monday, April 5, 2010?
4. In what ways did inspectors try to save the mine? Where did they fail?
5. Why do you think that the UBB mine disaster did not serve as a powerful focusing event with respect to the government's agenda in the way that the BP oil spill did?

NOTES

1. US Energy Information Administration (EIA), "Coal Explained: Where Our Coal Comes From" (Washington, DC: US Department of Energy, EIA, last updated April 24, 2017), www.eia.gov/energyexplained/index.cfm?page=coal_where (accessed June 14, 2017).

2. EIA, "Annual Coal Report: Table 1. Coal Production and Number of Mines by State and Mine Type, 2015 and 2014" (Washington, DC: US Department of Energy, EIA, 2015), www.eia.gov/coal/annual/pdf/table1.pdf.

3. EIA, "US Coal Production and Coal-Fired Electricity Generation Expected to Rise in Near Term" (Washington, DC: US Department of Energy, EIA, February 8, 2017), www.eia.gov/todayinenergy/detail.php?id=29872 (accessed June 13, 2017).

4. EIA, "International: Primary Coal Production 2014," www.eia.gov/beta /international/ (accessed June 13, 2017); see also EIA, "Coal Statistics" (Washington, DC: US Department of Energy, EIA, last updated April 3, 2017), www.eia.gov /energyexplained/index.cfm?page=coal_home#tab2 (accessed June 13, 2017).

5. EIA, "Today in Energy: Asia Leads Growth in Global Coal Production Since 1980" (Washington, DC: US Department of Energy, EIA, December 7, 2011), www .eia.gov/todayinenergy/detail.cfm?id=4210 (accessed August 5, 2013).

6. EIA, "Coal Explained: Use of Coal" (Washington, DC: US Department of Energy, EIA, May 30, 2017), www.eia.gov/energyexplained/index.cfm?page=coal_use (accessed June 13, 2017).

7. EIA, "Electricity Monthly Update: In 2016, Natural Gas Exceeds Coal for the First Time in the US Electricity Generation Mix" (Washington, DC: US Department of Energy, EIA, April 25, 2017), www.eia.gov/electricity/monthly/update /archive/april2017/ (accessed June 12, 2017).

8. Ibid.

9. EIA, "US Coal Exports and Imports Both Decline in 2016 as US Remains Net Coal Exporter" (Washington, DC: US Department of Energy, EIA, March 14, 2017), www.eia.gov/todayinenergy/detail.php?id=30332 (accessed June 13, 2017). See also EIA, "US Coal Exports Have Increased over the Past Six Months" (Washington, DC: US Department of Energy, EIA, July 18, 2017), www.eia.gov/todayinenergy /detail.php?id=32092 (accessed September 22, 2017).

10. EIA, "US Coal Production and Coal-Fired Electricity Generation Expected to Rise in Near Term" (February 8, 2017).

11. EIA, "Market Trends—Coal," in Annual Energy Outlook 2013 (Washington, DC: US Department of Energy, EIA, May 2, 2013), www.eia.gov/outlooks/archive/aeo13/.

12. EIA, "EIA Projects 28 Percent Increase in World Energy Use by 2040" (Washington, DC: US Department of Energy, EIA, September 14, 2017), www.eia.gov /todayinenergy/detail.php?id=32912 (accessed September 22, 2017).

13. Jack Fitzpatrick, "Coal Plants Are Shutting Down, With or Without the Clean Power Plan," Morning Consult, May 3, 2016, https://morningconsult.com/2016/05 /03/coal-plants-shutting-without-clean-power-plan/ (accessed June 14, 2017).

14. EIA, "Frequently Asked Questions: How Much of US Carbon Dioxide Emissions Are Associated with Electricity Generation?" (Washington, DC: US Department of Energy, EIA, May 10, 2017), www.eia.gov/tools/faqs/faq.php?id=77&t=3 (accessed September 22, 2017).

15. Rob Perks, "Appalachian Heartbreak: Time to End Mountaintop Removal Coal Mining" (New York: Natural Resources Defense Council, 2009), www.nrdc .org/land/appalachian/files/appalachian.pdf (accessed February 8, 2014).

16. EPA, Region 3 (Philadelphia, PA), *Mountaintop Mining/Valley Fills in Appalachia: Final Programmatic Environmental Impact Statement*, EPA 9-03-R-05002 (Washington, DC: EPA, October 5, 2005), 4, https://nepis.epa.gov/Exe/ZyPDF.cgi /20005XA6.PDF?Dockey=20005XA6.PDF.

17. MSHA, "From the Assistant Secretary's Desk, Commemorating the 35th Anniversary of the 1977 Mine Act—Historic Legislation That Saves Lives" (Washington, DC: US Department of Labor, MSHA, 2013), https://arlweb.msha.gov/FocusOn /35Years/35Years.asp; and Joseph Main, "The History of Mine Rescue" (Washington, DC: US Department of Labor, MSHA, undated), https://arlweb.msha.gov /TRAINING/LIBRARY/historyofminerescue/page6.asp (accessed September 22, 2017).

18. MSHA, "Celebrating 35 Years of the Mine Safety and Health Administration" (Washington, DC: US Department of Labor, MSHA, 2012), www.msha.gov /Focuson/35Years/MineActSuccess.asp (accessed February 26, 2014).

19. MSHA, "Fourth Quarter 2013 Mining Fatalities Report" (Washington, DC: US Department of Labor, MSHA, January 7, 2014); and MSHA, "Coal Mining Fatalities by State by Calendar Year" (Washington, DC: US Department of Labor, MSHA, 2017), https://arlweb.msha.gov/stats/charts/coalbystates.pdf (accessed September 22, 2017).

20. Peter A. Galuszka, *Thunder on the Mountain: Death at Massey and the Dirty Secrets Behind Big Coal* (New York: St. Martin's Press, 2012), 7.

21. WVMHST, "Report of Investigation into the Mine Explosion at the Upper Big Branch Mine in Boone/Raleigh Co., West Virginia" (Charleston, WV: WVMHST, February 23, 2012), section 3, 3, www.wvminesafety.org/PDFs/Performance /FINAL%20REPORT.pdf (accessed September 22, 2017).

22. J. Davitt McAteer, with Katie Beall, James A. Beck Jr., Patrick C. McGinley, Celeste Monforton, Deborah C. Roberts, Beth Spence, and Suzanne Weiss, *Upper Big Branch: The April 5, 2010, Explosion: A Failure of Basic Coal Mine Safety Practices: Report to the Governor* (Charleston, WV: Governor's Independent Investigation Panel, May 2011), 15, www.nttc.edu/ubb/ (accessed June 10, 2013).

23. WVMHST, "Report of Investigation into the Mine Explosion at the Upper Big Branch Mine," section 3, 2.

24. Carl Hoffman, "What Can Go Wrong: The Dangers in Longwall Mining," *Popular Mechanics*, April 8, 2010, www.popularmechanisms.com/science/energy /a5527/dangers-in-longwall-coal-mining/ (accessed November 2, 2017).

25. WVMHST, "Report of Investigation into the Mine Explosion at the Upper Big Branch Mine," 10.

26. McAteer et al., *Upper Big Branch: The April 5, 2010, Explosion*, 15.

27. Ibid., 19.

28. WVMHST, "Report of Investigation into the Mine Explosion at the Upper Big Branch Mine," section 4, 8.

29. McAteer et al., *Upper Big Branch: The April 5, 2010, Explosion*, 10.

30. UMWA president Cecil Roberts, cover letter, October 25, 2011, to UMWA, *Industrial Homicide: Report on the Upper Big Branch Mine Disaster: Report to the*

US Senate Subcommittee of the Committee on Appropriations (Columbus, OH: UMWA, May 20, 2010), 4.

31. McAteer et al., *Upper Big Branch: The April 5, 2010, Explosion*, 97.

32. Ibid., 96.

33. UMWA, *Industrial Homicide*, 74.

34. Jeff Goodell, "The Dark Lord of Coal Country," *Rolling Stone*, November 29, 2010, www.rollingstone.com/politics/news/the-dark-lord-of-coal-country-20101129 (accessed September 22, 2017).

35. Ibid.

36. Terri Day, "Buying Justice: Caperton v. A. T. Massey: Campaign Dollars, Mandatory Recusal, and Due Process," *Mississippi College Law Review* 28 (2009): 359–380.

37. Ibid., 360.

38. Adam Liptak, "Case May Alter Judge Elections Across Country," *New York Times*, February 15, 2009.

39. Ibid.

40. Alison Frankel, "Virginia Supreme Court Revives Epic Suit Against Massey Coal," *Reuters*, April 19, 2013.

41. Alexandra Zendrian, "The Cold Calculations of Coal Mining," *Forbes*, April 9, 2010 https://www.forbes.com/2010/04/09/coal-upper-big-branch-intelligent -investing-massey.html (accessed November 2, 2017).

42. "Blankenship Surfaces, Forms Kentucky Coal Company," *Charleston Gazette*, December 8, 2011.

43. UMWA, *Industrial Homicide*, 74.

44. Ken Ward Jr., "Miners Say Upper Big Branch Mine Cheated on Dust Sampling" (Washington, DC: Center for Public Integrity, July 8, 2012), www .publicintegrity.org/2012/07/08/9318/miners-say-upper-big-branch-mine-cheated -dust-sampling (accessed February 5, 2014).

45. McAteer, et al., *Upper Big Branch: The April 5, 2010, Explosion*, 32.

46. Ibid., 100.

47. Goodell, "The Dark Lord of Coal Country."

48. Dennis B. Roddy and Vivian Nereim, "A History of Violations at Upper Big Branch Mine," *Pittsburgh Press Gazette*, April 6, 2010.

49. McAteer et al., *Upper Big Branch: The April 5, 2010, Explosion*, 50.

50. Roddy and Nereim, "A History of Violations at Upper Big Branch Mine."

51. McAteer et al., *Upper Big Branch: The April 5, 2010, Explosion*, 54.

52. Ibid., 98.

53. Norman G. Page et al., "Appendix K: Inspection History," in *Report of Investigation: Fatal Underground Mine Explosion, April 5, 2010, Upper Big Branch Mine–South, Performance Coal Company, Montcoal, Raleigh County, West Virginia, ID No. 46-08436* (Washington, DC: US Department of Labor, MSHA, Coal Mine Safety and Health, 2011), https://arlweb.msha.gov/Fatals/2010/UBB/FTL10c0331noappx.pdf.

54. Galuszka, *Thunder on the Mountain*, 9.

55. Goodell, "The Dark Lord of Coal Country."

56. Giovanni Russonello, "The Coal Truth: Massey Was Worst Coal Company Even Before the April Disaster" (Washington, DC: American University, Investigative Reporting Workshop, November 23, 2010), http://investigativereportingworkshop .org/investigations/coal-truth/story/massey-had-worst-mine-fatality-record-even -april-d/ (accessed February 26, 2014).

57. McAteer et al., *Upper Big Branch: The April 5, 2010, Explosion*, 77.

58. Ibid.

59. Ibid., 83.

60. Russonello, "The Coal Truth."

61. Ibid.

62. Quoted in Goodell, "The Dark Lord of Coal Country."

63. Stephen Power, "Massey Hires Politically Connected PR Firm," *Wall Street Journal*, April 21, 2010.

64. Howard Berkes, "Mining CEO's Sudden Exit Follows Troubled Year," Special Series: Mine Safety in America, *National Public Radio*, December 4, 2010, www.npr .org/2010/12/04/131805599/mining-ceo-s-sudden-exit-follows-troubled-year?ft=1&f =1001 (accessed February 24, 2014).

65. Clifford Krauss, "Under Fire Since Explosion, Mining CEO Quits," *New York Times*, December 3, 2010.

66. McAteer et al., *Upper Big Branch: The April 5, 2010, Explosion*, 104.

67. Howard Berkes, "Massey Mine Execs Reap Millions in Takeover," *National Public Radio*, May 24, 2011, www.npr.org/blogs/thetwo-way/2011/05/24/136619832 /massey-mine-execs-reap-millions-in-takeover (accessed February 24, 2014).

68. Massey Energy Company, *Preliminary Report of Investigation Upper Big Branch Mining Explosion*, 2011, 2, www.eenews.net/assets/2011/06/03/document _pm_01.pdf.

69. Taylor Kuykendall, "Massey Issues Report on Its Investigation," *(Beckley, West Virginia) Register-Herald*, June 4, 2011.

70. Stephanie McCrummen, "After Massey Mine Disaster Killed Their Son, Settlement of Millions Is Worth Little," *Washington Post*, July 12, 2012.

71. Ibid.

72. McAteer et al., *Upper Big Branch: The April 5, 2010, Explosion*, 4.

73. Ibid., 108.

74. Ibid., 92.

75. Dave Jamieson, "Upper Big Branch Report: Feds Release Devastating Findings on Preventable Mine Disaster," *Huffington Post*, December 6, 2011, www .huffingtonpost.com/2011/12/06/upper-big-branch-report-findings_n_1132462 .html (accessed February 7, 2014).

76. Page et al., *Report of Investigation*, 2.

77. Ibid., 5.

78. WVMHST, "Report of Investigation into the Mine Explosion at the Upper Big Branch Mine," 5.

79. Ibid., 5 (sect. 6).

80. Ibid., 3.

81. CNN Library, "US Mine Disasters Fast Facts," *CNN*, July 13, 2013, www.cnn.com /2013/07/13/us/u-s-mine-disasters-fast-facts/index.html (accessed October 10, 2013).

82. Ken Ward Jr., "UBB Investigation Continues, Goodwin Says," *Charleston Gazette*, February 16, 2014.

83. Cathleen Moxley and Anna Baxter, "Ex W.Va. Mine Boss Says Lawyer Made Him a Scapegoat," *WSAZ News Channel 3*, September 24, 2013, www.wsaz.com /home/headlines/UBB_Mine_Boss_Charged_With_Fraud__139969353.html (accessed February 7, 2014).

84. "David Hughart Sentenced to Nearly 4 Years in Prison," *State Journal*, September 10, 2013.

85. Ward, "UBB Investigation Continues, Goodwin Says."

86. John Raby, "US Mine Blast: Ex-Coal CEO Blankenship at End of Prison Term," Associated Press, May 10, 2017.

87. "Don Blankenship Resurfaces, Forms Kentucky Coal Company," *WSAZ News Channel 3*, December 8, 2011, www.wsaz.com/news/headlines/Don_Blankenship _Resurfaces_Forms_Ky_Coal_Company__135259468.html (accessed February 26, 2014).

88. Goodell, "The Dark Lord of Coal Country."

89. Ian Urbina, "Authorities Vow to Close Mines Found to Be Unsafe," *New York Times*, April 27, 2010.

90. Associated Press, "Two Workers Are Killed in Kentucky Mine Collapse," *New York Times*, April 29, 2010.

91. Howard Berkes and Robert Benincasa, "Labor Dept. Asks Court to Close Massey Mine in KY," *National Public Radio*, November 3, 2010, www.npr.org/2010 /11/03/130596700/labor-dept-asks-court-to-close-massey-mine-in-ky (accessed February 26, 2014).

92. Jim Morris, "Six Questions for J. Davitt McAteer," Center for Public Integrity, April 5, 2011, www.publicintegrity.org/2011/04/05/3929/six-questions-j-davitt -mcateer (accessed February 5, 2014).

93. US House of Representatives, Committee on Education and the Workforce, "Learning from the Upper Big Branch Tragedy" (statement by Representative John Kline), March 27, 2012.

94. Patrick Reis (of Greenwire), "West Virginia Sues Obama, EPA over Mining Coal Regulations," *New York Times*, October 6, 2010.

95. "Upper Big Branch Mine to Be Permanently Sealed," *CNN*, April 5, 2012, www .cnn.com/2012/04/05/us/west-virginia-mine-disaster/ (accessed October 10, 2013).

96. WVMHST, "Report of Investigation into the Mine Explosion at the Upper Big Branch Mine."

97. Goodell, "The Dark Lord of Coal Country."

98. Ibid.

CHAPTER 5

The Town That Became a Superfund Site

ASBESTOS IN LIBBY, MONTANA

The stories of Union Carbide in Bhopal, India, the *Deepwater Horizon* in the Gulf of Mexico, and the Upper Big Branch mine in West Virginia share one sad commonality: all three involve a violent explosion that resulted in immediate loss of life. The story in this chapter has no such dramatic climax. No explosions lit the night, and no sirens broke the silence. Instead, disaster came gradually, unfolding over decades. Nevertheless, it is a tragic story second to the Bhopal story in loss of life. For in the small town of Libby, Montana, entire families were lost—and continue to be lost—to the ravages of asbestos-related diseases, including asbestosis, mesothelioma, and lung cancer.

This is a tale of a sinister invader that came in the form of the tiny asbestos fibers in the vermiculite ore mined from the Zonolite mine owned by W. R. Grace and Company. Unlike most instances of occupational exposure to toxic substances, it was not workers alone who were at risk in Libby, Montana, and nearby Troy; asbestos lingered in the air for all residents of these towns to breathe. These fibers found their way into the lungs of the family members who greeted workers returning home from the mine. A welcome-home hug or a worker shrugging off a work jacket could send asbestos into the air to be breathed by anyone nearby. Equally troubling, Libby residents who had no connection at all with the vermiculite mine were also at risk from this deadly intruder. Vermiculite was used to insulate attics and condition soils in Libby gardens. Children played in a pile of vermiculite next to the athletic field. For more than seven decades, the widespread contamination of Libby continued unabated, and the result was one of the largest environmental exposures to asbestos in US history.

Libby was the epicenter of this tragedy, but not the only location of victims, for the Zonolite vermiculite mine was a large operation. While in business, the Libby mine produced an estimated 80 percent of the world's supply of vermiculite.[1] Asbestos-laced vermiculite ore was sent to more than three hundred processing plants across the United States, where it was processed, bagged, and delivered to thousands of homes for attic and wall insulation. Thousands of plant workers were exposed to asbestos fibers, and potentially hundreds of thousands of people who purchased the Zonolite vermiculite product have been, or will be, exposed. When the World Trade Center in New York collapsed on September 11, 2001, asbestos from Libby was part of the toxic cocktail released into the air.[2]

Although some asbestos-containing materials have been banned in this country, others have not, despite the harm done in Libby and the massive number of wrongful death lawsuits brought against asbestos manufacturers like the Johns Manville Corporation.[3] Even if no towns suffer in the future as much as Libby has, countless people have already been exposed to asbestos or may be exposed in the future. Asbestos will undoubtedly claim more lives, in Libby and throughout the world.

In short, this is a story of a company that mined and distributed vermiculite laced with asbestos knowing that the consequences were potentially deadly and that a single town was exceptionally exposed to this danger. This chapter, like the previous three, explores the reasons for what, in this case, was a slow-moving disaster. After providing some background about asbestos, the chapter explores the reasons behind the Libby disaster, detailing how a growing company allowed its workers and the town to be exposed to asbestos for decades.

Although the story has its share of villains, it also has many heroes, including two in particular: the investigative reporter Andrew Schneider and Libby resident Gayla Benefield, both of whom helped push Libby onto the national agenda. It was not only the disaster but their deeds that spurred government action. The chapter closes with a discussion of the legal ramifications for W. R. Grace stemming from its role in the disaster.

ASBESTOS: HISTORICAL USES AND HEALTH RISKS

Derived from a Greek word meaning "inextinguishable" or "unquenchable," asbestos has long been heralded for its unique fireproofing and insulating properties. Ancient cultures recognized this amazing property of asbestos. Greeks wove it into textiles, and Romans used it in building materials. Dubbed in modern times a "magic mineral" and a "miracle fiber," asbestos was thought to be an ideal insulating, fireproofing, or acoustical sound-proofing material for the industrial age, and its use was widespread.

Asbestos is a mineral fiber that naturally occurs in rocks and soil. There are six types of asbestos: chrysotile, amosite, actinolite, anthophyllite, crocidolite, and tremolite. These six types belong to two major groups of asbestos minerals: serpentine and amphibole.[4] Chrysotile asbestos is the most widely used in commercial applications; tremolite asbestos (the kind found in the vermiculite mined in Libby) belongs to the second group (amphibole). The straight, needlelike fibers of tremolite asbestos have been shown to carry more potent cancer-causing potential.[5] The health risks of asbestos depend on its degree of friability—that is, the extent to which it can be easily crumbled by hand and released into the air.[6]

Asbestos poses the greatest health risk when friable fibers are in the air because their small size allows them to be inhaled deeply into the lungs. Once inhaled, the fibers become trapped in the lungs and the body cannot break down or eliminate them.[7] The risk for asbestos-related disease depends on the type of asbestos fibers inhaled, the level and duration of exposure, and whether or not the exposed individual smokes. Each exposure increases the likelihood of developing asbestos-related disease, but even infrequent exposures to asbestos pose serious risks.[8] Once asbestos fibers are breathed into the lungs, the latency period between exposure and the onset of asbestos-related disease is typically ten to twenty-five years.

Asbestos exposure is primarily associated with three diseases: asbestosis, lung cancer, and mesothelioma. These diseases occur as asbestos fibers settle into the lower lobes of the lungs, causing irritation, inflammation, and ultimately calcification. Asbestosis results in a serious scarring of the lung tissue (fibrosis) that makes breathing difficult, prompting shortness of breath and a chronic cough. Extended exposure to asbestos leads to an accumulation of fibers in lung tissue, which sets the stage for lung tissue to thicken, causing pain. Although asbestosis is not cancer, it is often fatal. People who suffer from asbestosis also have an eight to ten times higher risk of developing lung cancer.[9]

Asbestos is classified as a human carcinogen. Lung cancer accounts for about half of all asbestos-related disease.[10] Asbestos workers are about five times more likely to develop lung cancer than workers not exposed to asbestos in the workplace, and asbestos workers who smoke are fifty to ninety times more likely to develop lung cancer than nonsmokers.[11] Mesothelioma, a rare and frequently lethal cancer that most often affects the thin membranes lining organs in the chest (pleura) and abdomen (peritoneum), is closely linked to asbestos exposure, especially amphibole asbestos (the type found in vermiculite).[12] The National Cancer Institute estimates that the five-year survival rate for mesothelioma is between 5 and 10 percent.[13] Mesothelioma accounts for roughly 20 percent of deaths due to asbestos exposure. Mesothelioma rarely occurs without exposure to asbestos, which is why it is often called the "asbestos cancer."[14] Studies have also looked at elevated risk for other types of cancer linked to swallowing asbestos fibers, such as cancers of the throat, stomach, and colon.[15]

A History of Asbestos Use in the United States

The United States was quick to recognize the commercial potential of asbestos, which is easy to manipulate, yet strong and fire-resistant. The production of asbestos-containing materials was championed through the last half of the nineteenth century and first half of the twentieth by people inside and outside of the national government. Asbestos became a major staple of US manufacturing and had an estimated three thousand industrial applications. Automakers used it in brake pads and clutch plates. Pipe wrapping, roofing, wall and ceiling insulation, siding, flooring, and the insides of boilers all contained asbestos.

No company realized the potential of asbestos more than Johns Manville. In 1858, H. W. Johns Manufacturing Company in New York (later to become Johns Manville) opened for business to provide asbestos as a fire-resistant roofing material.[16] The company soon saw the value of asbestos in the emerging automobile industry and created "non-burn" asbestos brake linings as early as 1916. Soon after, Manville Covering Company, the second half of the early Johns Manville Company, opened in Milwaukee, Wisconsin, to provide asbestos as a heat-insulating material.[17] As the company slogan went, "When you think of asbestos, you think of Johns-Manville."[18] The company went public in 1927, and three years later it was selected to join the Dow Jones Industrials, owing in part to the promising future of asbestos. Ironically, the first paper on asbestos-related disease had been published just a few years prior. The word "asbestosis" was first used in an article in the *British Medical Journal* in 1924 in reference to the cause of death for a young woman who worked in an asbestos factory.[19]

The federal government supported the asbestos industry despite knowing of the potential health risks. The government suspected that asbestos was deadly as early as 1918, yet did very little to protect workers. That year, the US Department of Labor Statistics published a report noting that workers exposed to asbestos were experiencing early deaths. During World War II, asbestos insulation was used extensively on US Navy ships, even though the Navy's surgeon general had issued warnings about asbestos in 1939.[20] Two years later, Commander C. S. Stephenson, the Navy's chief officer for preventative medicine, warned that the Navy was not protecting shipyard workers from exposure to asbestos.

But the warnings went unheeded. Asbestos was needed as part of the war effort. In 1942 alone, the War Production Board allocated 40 percent of the estimated 36.8 million pounds of available asbestos-containing pipe insulation to the nation's shipyards, exposing thousands of shipyard workers.[21] Decades later, the journalist Bill Burke observed:

> Working in an American shipyard during World War II would prove to be almost as deadly as fighting in the war. The combat death rate was about 18 per thousand armed service members. For every thousand wartime shipyard employees, about 14 died of asbestos-related cancer, and an unknown number died of asbestosis or complications from it.[22]

Even after World War II, asbestos use continued. It was not until 1971 that the Occupational Safety and Health Administration (OSHA) promulgated the first federal regulation for occupational exposure to asbestos. The Navy would not ban asbestos use on new ships until 1973. The EPA also began regulating asbestos as a hazardous air pollutant in the 1970s. By that time, millions of people had been exposed.

Asbestos Exposure Leads to Litigation

The story of asbestos exposure around the country provides a chilling backdrop to the Libby disaster. Beginning in the mid-1960s, thousands of people began developing serious illnesses as a result of working with asbestos. Experts estimate that between 1940 and 1980, 27 million Americans experienced significant occupational exposure to asbestos.[23] The results were deadly: between 1980 and 1995, an estimated 149,350 people in the United

States died of occupational asbestos disease. That number surpassed the combined total of all other workplace injuries and illnesses for that period: 140,365.[24] Due to the long latency periods for asbestos-related diseases and the continued presence in the built environment of asbestos-containing materials, deaths will continue into the future. More recently, it is estimated that 10,000 asbestos-related deaths occur annually in this country, and that the rate of asbestos-related deaths will not decline until sometime after 2020.[25] The Centers for Disease Control and Prevention (CDC) found that 45,221 people died of mesothelioma between 1999 and 2015, and that the persistent rate of mesothelioma deaths indicates that asbestos exposure is still substantial.[26] Suffice it to say that breathing asbestos fibers poses serious, even fatal, consequences that may not materialize for decades as permanently lodged asbestos fibers continuously assault the lungs.

The dramatic number of asbestos-related deaths was not lost on the media or litigators. Stories in national newspapers and on television news noting the link between asbestos exposure and premature deaths began to appear in the mid-1960s. The first asbestos products lawsuit was filed in 1966, and the first against Johns Mansville came in 1974. Media attention escalated over the next decade, and over one hundred stories noting the dangers of asbestos were published between 1982 and 1985. This drew the attention of the public and opened the possibility of large class action lawsuits.

Under an onslaught of asbestos injury claims and lawsuits, Johns Manville filed for bankruptcy in 1982. At the time, it was the largest bankruptcy in US history. As a condition of its emergence from bankruptcy, the company was forced to create the Manville Personal Injury Settlement Trust in 1988, with an initial deposit for claimants of $2.5 billion. (Johns Mansville was acquired by Berkshire Hathaway in 2001.) As of 2012, the trust had paid $4.3 billion and settled 773,990 claims.[27] Other asbestos companies followed this major asbestos manufacturer into bankruptcy protection and created similar trust funds. As of 2016, at least sixty such trust funds had been established.[28] When all asbestos trusts are considered, more than $17.5 billion has been paid to settle over three million claims.[29]

Thus began what would be known as the three waves of asbestos litigation: the first, from workers at asbestos mines and factories; the second from workers injured by exposure at sites where asbestos-containing materials were processed or used (such as shipyard workers); and the third from construction workers and others injured by the installation or removal of asbestos-containing materials, such as drywall and ceiling or floor tiles. Still others would suffer from non-occupational exposure—a major factor in the Libby story. The next section highlights the link between vermiculite and asbestos and the rise to prominence of the Zonolite Mining Company.

THE HISTORY OF THE ZONOLITE MINE

The story of Libby, Montana, is a story about asbestos, not because asbestos was mined directly, but rather because asbestos was in the vermiculite mined on Zonolite Mountain. The value of vermiculite was not commercially known until 1916, when Edgar Alley discovered the mineral inside of Zonolite Mountain near Libby. Alley found that vermiculite expands when heated, a process called "exfoliation" or "popping." The process formed a lightweight material that not only was a superior insulator but was also easy to use and inexpensive. In 1919, Alley named this popped vermiculite "Zonolite." It took little time for several large manufacturers, including Johns Mansville, to recognize the worth of this lightweight mineral. Commercial mining of vermiculite began in 1923, and the Zonolite Company was launched.[30] Vermiculite soon became widely used as an insulating material, in packing materials, and as a soil amendment. Within ten years, Alley's popped vermiculite was in high demand, and a plant had been built that could process 110 tons of vermiculite per day.[31]

In 1939, the Zonolite operation merged with a competitor on the mountain, Universal Insulation Company, and the Universal Zonolite Company was born. By 1940, more than 80 percent of the country's supply of vermiculite came from Libby, Montana.[32] For a time, it appeared that Libby had been blessed: because it was near Zonolite Mountain, which contained the largest vermiculite deposit in the United States, the town was home to the world's largest vermiculite mining operation, which produced 150,000 tons per year.[33]

But the town's euphoria over having a robust industry to support the local economy began to erode as adverse health issues at the Zonolite mine became part of the story. Signs that conditions at the mine could have deadly consequences for its workers began in 1944, when a state inspector warned mining officials that they needed to install dust control equipment and recommended that workers be given respirators to avoid inhaling "nuisance dust." In 1956, Ben Wake, an industrial hygiene engineer for Montana's Health Department, inspected the mine. He warned Zonolite managers that the dust was more than just bothersome—it was dangerous. His four-page report revealed that "asbestos in the air is of considerable toxicity" and that the company had "poor policy in matters of maintenance and operation of the plant."[34]

Three years later, in 1959, Wake found that the changes he recommended had not been made. This was the same year an official diagnosis of asbestosis was first given to a Zonolite miner. It was also the year Montana passed an occupational disease law that held companies responsible for injuries and illness at the workplace.

In response to the law, Zonolite ordered all its employees to have chest X-rays taken at a local hospital in order to establish how many were already sick from exposure to asbestos coming from the vermiculite mine. This was not a moral gesture. Executives reasoned that it made good business sense, for the new state law was not retroactively applied to existing conditions. Thus, wrote plant manager Earl Lovick, X-rays were necessary "in order to protect ourselves and place ourselves on record as to the condition of our employees as of the effective date of the law."[35] The record was not a good one: more than half of the chest X-rays (82 out of 130) showed symptoms of lung disease, and over one-third of the X-rays showed lung abnormalities or signs of asbestosis.

Inspector Wake sampled the mine a year later, in 1960. Once again, asbestos was in the dust coming from the mining process and the Zonolite Company was doing little about it. A year later, air sampling at the Libby mine by the Montana State Board of Health Survey showed extremely high concentrations of asbestos dust, which was identified as coming from tremolite asbestos, the most dangerous type of asbestos. In 1962, Wake sent samples of the dust to the federal Public Health Service to test for asbestos levels. What came back surprised him: asbestos levels in the mine had continued to climb. According to the report, the samples were composed of a staggering 40 percent asbestos. He passed this finding on to the company in his next inspection report, noting that "no progress has been made in reducing dust concentrations in the dry mill to an acceptable level and that, indeed, the dust concentrations had increased substantially over those in the past."[36]

In the meantime, additional workers received medical evaluations concluding that their lung problems were almost certainly from asbestos in the dust. Such evaluations, however, were not taken as reason to shutter the mine or milling operation, or even to slow it down. Instead, the workers' medical evaluations prompted Lovick to write to one of the company's largest customers, Western Mineral Products Company, to warn it about asbestos: "Asbestos is a cause of asbestosis, which has been a matter of concern. There is a relatively large amount of asbestos dust present in our mill, and this is difficult to control."[37] Workers, however, were not informed of the results. This company nondisclosure practice would continue after W. R. Grace bought the Zonolite mine in 1963.

THE LIBBY DISASTER: THE RESULT OF THE W. R. GRACE COMPANY CULTURE

Two of the four common factors in the stories of industrial disasters identified in Chapter 1 are a disregard for environmental and safety standards, and a pursuit of profits at all costs. In this story, the company's culture of complacency around environmental and safety protocols was closely connected to its focus on the bottom line. Another characteristic that the Libby disaster had in common with other industrial disasters was the company's lack of planning and preparation. In this section, we look at these factors in greater depth.

Complacency and the Pursuit of Profits

W. R. Grace bought the Zonolite mining operation knowing the details of the asbestos exposures in the mine and milling operations through the inspection reports and the X-ray records (though the company would later dispute this fact). Instead of making necessary changes, the company continued the practices of its predecessor in doing next to nothing to protect its workers.

As he had with Zonolite managers, Wake sounded the alarm in 1964 in his reports to W. R. Grace managers. He hoped, perhaps, that the new owners would respond differently. However, little changed. For example, W. R. Grace refused to engage in the kind of basic housekeeping solutions that might have minimized exposure. One of the most unpleasant jobs was working as a sweeper at the dry mill, where the vermiculite was processed. Dust accumulated on ceilings, walls, and light fixtures. Men used a push broom to clean up this asbestos-laden dust, sending plumes of it into the air—which they breathed every day. The dry mill was six stories tall. Men would start at the top floor, sweeping the dust they could get at into bins; some of the dust would work its way through holes in the floor into the floor below. As former employee Les Skramstad would describe it, "Probably that's the worst place you could ever imagine being. If they had hundreds of vacuums blowing all the dirt in town into this room, that might be comparable . . . that dust stuck to everything. You couldn't see the light bulbs. . . . When we'd get it all down to the last floor, it would be three feet deep."[38]

W. R. Grace provided respirators for the workers, but the mill operation was so filled with dust that the respirators would clog. Because they believed it was only nuisance dust, the men would take off the clogged respirators to be able to work more efficiently.[39] Had they known that the dust was heavily laced with asbestos, they might have kept their respirators on. In any event, respirators were no substitute for good maintenance, a better, cleaner system, or shutting the mill until the dust was completely cleared.

W. R. Grace installed a ventilation fan at the dry mill to exhaust the dust outside of the mill. But that only expanded the scope of exposure. Now the dust blew out from the mill onto the shop yard, where maintenance workers were stationed. It would be four more years before W. R. Grace installed a vent shaft to send the dust away from maintenance workers. The dry mill operated until 1974, when a new wet mill came on-line, partly in response to a state citation for air pollution violations.

Steadfast and determined, Wake filed inspection reports for over a decade in which he noted the "considerable toxicity" of the asbestos-laden dust. All reports were marked "confidential" and sent to W. R. Grace managers. However, executives largely ignored the inspection reports, choosing instead to file them away. The result: workers were not informed about the dangers they faced every day. Some thirty years later, however, the reports provided sobering evidence that W. R. Grace officials knew they were exposing workers to lethal exposures and did nothing.

Well, not quite nothing. Following Zonolite's lead, W. R. Grace officials tracked the progress of asbestos disease in their workers. In 1964, W. R. Grace initiated annual X-ray testing of all its employees. The results were just as sobering as the results of the tests conducted by Zonolite: X-rays consistently showed that one in four workers had a lung condition suggesting the start or existence of asbestos-related disease. For example, a confidential study conducted in 1969 by W. R. Grace of its Libby employees revealed that 17 percent of workers with fewer than five years at the mine were likely to have lung disease, 45 percent of workers with between eleven and twenty years of service showed lung disease, and an astounding 65 percent of workers with twenty-one to twenty-five years of service had lung disease.[40] To put this another way, anyone who worked long enough to retire from W. R. Grace's vermiculite mine was likely to leave with a death sentence as a consequence of asbestos exposure.

The X-ray tests became a source of internal data for the company, which developed lists of employees that described the state of the disease by worker name. Yet W. R. Grace still did not inform workers of the results of the tests, did not share with those who had asbestosis the progression over time of their disease, and told none of their workers that they saw evidence of lung disease in their X-rays. W. R. Grace's insurance carrier, Maryland Casualty Company, chastised the company, saying that failure to warn workers of the illness shown in their X-rays was "not humane and in direct violation of federal law."[41]

The company also conducted its own asbestos toxicity tests in laboratory animals. These tests confirmed what the company knew—exposure to tremolite fibers produced lung disease and cancerous tumors. Still W. R. Grace did not warn its workers, even in the face of mounting evidence that asbestos would take a very high toll in human life. Miners would not be told until 1979 about the danger of asbestos in the dust—twenty-three years after Wake issued his first report, and fifteen years after W. R. Grace required its first X-rays of employees. Even then, the news broke only indirectly, during an inspection by the US Mine Safety and Health Administration (MSHA). As he talked with the federal inspector, union president Bob Wilkins recalled, he commented, "We're fortunate that we don't have an asbestos problem, we have tremolite."[42] After the inspector showed Wilkins that tremolite was indeed asbestos, Wilkins went to Lovick, the plant manager, who confirmed that this was true.

As found in the other stories in Chapters 2 to 4, W. R. Grace maintained a razor-sharp focus on the bottom line. One example of this attachment to profits at all costs came in 1977, when workers asked for additional showers and uniforms so that men could leave the dust at the plant and not take it home. At the time, there was only one shower for a crew of 140 men. It would be five years before W. R. Grace managers made a decision regarding the request. In 1982, they balked at the anticipated costs ($373,000) for the showers, uniforms, and overtime costs that might be paid if the men showered after their shift. Ironically, 1982 was an excellent year for the company, which set a record for profits of nearly $131 million.[43] Years would go by before W. R. Grace issued coveralls to employees to help reduce the amount of dust carried on clothing.[44]

Perhaps the best example of pursuing profit while sacrificing the health of workers was W. R. Grace's human resources policy. After discovering such high numbers of men affected with asbestos-related disease, W. R. Grace began shifting men who became ill to less hazardous jobs. This was not, however, an attempt to prolong their lives or make them more comfortable. It was instead an effort to avoid paying for disability care, as reflected in a memo to management by W. R. Grace's safety officer, Peter Kostic.

He recommended keeping sick workers away from the dust to increase the chances that they would stay on the job, thus "precluding the high cost of total disability."[45]

W. R. Grace failed workers in its Libby facility, but it also failed vermiculite ore–processing plants across the country. When R. M. Vinning, president of the W. R. Grace Construction Products Group, wrote in 1969 to J. Peter Grace, the company president, warning him that tremolite was a health hazard that could not be separated from the ore shipped from Libby, the president responded that vermiculite safety concerns would be dealt with "singly, as they are forced to comply, and to buy as much time as possible."[46] In 1971, W. R. Grace initiated annual X-ray testing in all vermiculite expanding plants after reports of sickened workers. The EPA would later investigate over two hundred facilities across the country that processed Libby ore as part of its effort to characterize asbestos exposure.

If W. R. Grace failed its employees, it also failed the town of Libby. As early as 1965, W. R. Grace knew that asbestos was in the town, brought in by the dust from the mine. An internal W. R. Grace memo revealed during trial noted that the company's air monitoring system detected asbestos in downtown Libby on many dry days. On a still day, most of the dust settled around the mill. But if the wind blew from the east, a film of white dust covered the town. Tests by W. R. Grace found that twenty-four thousand pounds of dust a day were expelled from the large stack at the company's dry mill, and that this dust had an average asbestos content of 20 percent. On windy days, Libby could be coated with roughly five thousand pounds of asbestos. It would be eleven years before W. R. Grace replaced the lethal dry mill, with its huge exhaust fan, with a cleaner "wet mill" where water sprayed on the mined vermiculite would help reduce airborne dust.

Dust carried on the wind was but one way the town was exposed. Vermiculite mining is a dusty occupation. Since workers weren't told that the dust they carried on their clothing was lethal, it became part of many Libby homes. Children played on the floors and breathed asbestos fibers; homemakers sorted laundry or vacuumed, breathing asbestos fibers that lingered in the air. They used the readily available vermiculite to condition their garden soil and insulate their homes. W. R. Grace permitted workers and Libby residents to take the "waste" vermiculite—vermiculite that was not up to commercial standards—by the pickup load for personal use. W. R. Grace even provided some of this waste vermiculite for the school's running track and the local ice rink. These pernicious practices allowed the community to become permeated with asbestos. It is sadly not surprising that the EPA would find so much contamination in Libby when the agency began sampling in 2000.

In 1990, W. R. Grace closed the mine. By that time, attitudes toward asbestos had changed. The national market for asbestos had plummeted amid escalating asbestos injury lawsuits. The year before, the EPA had banned most asbestos-containing products under authority provided in Section 6 of the Toxic Substances Control Act. Though the rule was vacated in 1991 by the Fifth Circuit Court of Appeals, most asbestos-containing materials were increasingly unpopular, and some materials remained banned. W. R. Grace was quickly swept up in the asbestos litigation tsunami, and its Libby plant was no longer a profitable venture.

Inadequate Planning and Preparation

It appears that W. R. Grace undertook very little planning to provide adequate protection from asbestos exposure for workers or the environment. The company had a safety committee and did act to correct issues brought forward during inspections. However, that practice did not include an adequate response to the most terrifying safety condition: exposure to asbestos. The company allowed horrifying conditions to persist at the dry mill from the time it acquired the Zonolite operations until 1974, when being cited for air pollution violations triggered the company's construction of a safer wet mill.

As described previously, the company had no plan in place to inform workers about the dust around the mine and milling operations. Men were needlessly exposed for years, their lives cut short because of the company's unwillingness to engage in safe practices. Nor did the company try to protect the town from asbestos, even after its executives knew it was in the air. Children played in waste vermiculite, some of it located by the town's baseball field, and the company did not stop them. W. R. Grace had no plan in place to limit asbestos exposure wherever it occurred, just as Union Carbide had no plan for dealing with community exposure to the deadly methyl isocyanate gas.

This is perhaps best illustrated through the example of the plant manager, Earl Lovick. Lovick managed the plant for the Zonolite Company and stayed on as manager for W. R. Grace. He knew from at least 1959 that asbestos was part of the vermiculite ore, as he had access to all the studies and inspection reports. He knew every place at the facility where high levels of asbestos were present. After the X-ray testing began, he knew the staggering rate at which his employees were getting lung disease. When testifying during a 1997 court case, Lovick was asked why he did not require that respirators be worn, or whether he ever told the men why respirators were essential safety equipment. His response was that there are some things you shouldn't need to explain. "You shouldn't have to tell an employee [not to] put their fingers on a red hot iron, either, but we never told them that."[47] Whether or not Lovick fits the description of a villain, he certainly should be faulted for taking a cavalier attitude toward preparing workers or planning for this disaster. Lovick died of lung cancer in 1997.

INEFFECTIVE REGULATORY OVERSIGHT

As in the previous three industrial disaster stories, government oversight of the unfolding disaster in Libby was ineffective. Montana may be judged to have failed Libby workers and their families in part because of the importance to the state of the mining industry, which has a long tradition there; the state has often been reluctant to impose strict regulations on the industry. The state Department of Health performed inspections from 1956 until the mine closed in 1990. State inspections found unsanitary and unhealthful conditions, and state inspectors, especially Ben Wake, knew that the "nuisance" dust contained tremolite asbestos. The state notified Zonolite, and later W. R. Grace, of the dangerous conditions and the seriousness of asbestos-related disease, but it did not inform mine workers about the dangers of asbestos. State regulations required that only company officials be told, and reports were never disclosed publicly.

The state's long-standing relationship with the mining industry was often one of accommodation—not worker protection. In 1974, a state engineer found that one-third of the samples taken at the new wet mill still exceeded existing limits for asbestos. Yet, instead of insisting on immediate improvements, he wrote, "We are now satisfied that the new process and plant have significantly improved working conditions. At some time in the future ... I would like to verify that employee exposures to asbestos remain acceptable."[48] Years would pass before the next state inspection of the Zonolite mine operations. Even then, the tone was one of accommodation. In 1984, for example, Bob Raish, a Department of Health supervisor, wrote to Lovick to tell him to anticipate an inspection. W. R. Grace responded that it would send its own air samples, though the company never supplied its data.[49] Between 1979 and the cessation of mining operations in 1990, the state performed fourteen inspections—more than one a year—and two post-closure inspections.[50] But Montana, through its inspections and laws, never closed the vermiculite operation. Other than telling Zonolite, and later W. R. Grace, managers to correct problems, the state of Montana took few steps to ensure a safe working environment.

In 2004, the Supreme Court of Montana found in *Orr v. State of Montana* that the state had failed its legal obligation to protect the workers, noting in its decision:

> The State argues that it could not foresee that the Mine owner would not fulfill its legal obligations as landowner and employer. This rings hollow in light of obvious and objective indications that neither Zonolite nor Grace was protecting its employees. Plainly, the State knew as a result of its inspections that the Mine's owner was doing nothing to protect the workers from the toxins in their midst.[51]

That decision created an opportunity for Libby residents and former W. R. Grace workers to file lawsuits against the state of Montana. In 2011, Montana paid $43 million to settle a claim from more than 1,300 plaintiffs. A second major settlement was made in 2017, when the state paid an additional $25 million to over 1,000 claimants. Montana agencies argued unsuccessfully that the state had no legal obligation to provide warning of the mine's dangers.[52]

Efforts by federal regulators also fell short. The first federal inspection by the US Bureau of Mines in 1971 revealed what others had known: unacceptably high levels of asbestos exposure. MSHA took over regulatory responsibilities in 1978. Three factors hampered the ability of MSHA inspectors to force W. R. Grace to change its practices: First, W. R. Grace had advance notice of impending inspections; it appears that local residents would warn the company when federal inspectors checked into a Libby motel. Second, lab analysis of the samples taken at the facility typically met the federal exposure limit for asbestos, but the method of analysis prevalent at the time was not adequate to measure the extent of asbestos. Finally, MSHA inspectors had no authority to monitor or control the asbestos that was in the town.

When the story broke about Libby, the MSHA administrator, Davitt McAteer, assumed responsibility for failing to protect W. R. Grace workers. (You may recall the name from the previous chapter.) He initiated special inspections at all mines in the country where asbestos exposure might threaten miners, and he employed different testing methods, which included having miners wear portable air pumps. These data would now better inform the inspection process. Of course, it was too late for Libby. How this story garnered national attention is the subject of the next section.

THE GOVERNMENT RESPONDS: AGENDA-SETTING AND A MASSIVE CLEANUP

As described in Chapter 1, focusing events can trigger governmental action and media attention. Such was the case for Libby. In the summer of 1999, Andrew Schneider, a senior investigative reporter for the *Seattle Post-Intelligencer,* along with two other writers and photographers from the newspaper, was hard at work in Montana. The reporters were putting the final touches on a major series exposing the long-standing consequences of the 1872 General Mining Law on public lands in the West.[53] Montana was the last of nine states they had investigated and the road-weary staff was ready to go home when they got a tip about strange goings-on at a vermiculite mine near a small town called Libby. Schneider and photographer Gilbert Arias decided to explore the lead, beginning with a law firm revealed in the tip.[54] The lawyer, Roger Sullivan, was tight-lipped about his pending cases, but did confirm that people had died because of contamination at the mine.

That was enough to convince Schneider to head to Libby and see the mine for himself. Once there, he met with Gayla Benefield, a longtime resident of Libby, who took them to the shuttered Zonolite mining operation. The Zonolite mine had been closed since 1990. W. R. Grace was asking Montana's Department of Environmental Quality to return the last of its reclamation bond money, a request that had triggered the ire of Benefield and other residents. According to the company, the land had been reclaimed and was now covered with green grass and pine seedlings. Instead, a towering tailing pile loomed at the site, containing rocks, soil, and dust. Lots of dust. Dust that was laden with asbestos.

At first, Schneider was reluctant to pursue the story.[55] He did not see anything especially newsworthy in a story about poor mine reclamation. Sadly, failure to reclaim mining sites was nothing new: abandoned or partially reclaimed sites bigger than the one at the Zonolite mine dotted the country, especially in the West. What moved him into an investigative frame of mind was Benefield's comment that the mine had killed both her mother and father. "Dad came home covered in white dust and it was all over the place. Mom was a fanatic about keeping the house clean. She bought the newest Kirby vacuum, but it would just suck in the fibers and blow them right back into the air."[56] Benefield went on: "Lots of people here have died because of that mine. And a hell of a lot more are dying and no one cares. . . . Not one single government agency. Not here. Not in Helena. Not in Washington."[57]

Photo 5.1: Gayla Benefield holds the cowboy hat of Les Skramstad, friend and mine worker who died of asbestosis, at the Russell Smith Federal District Courthouse in Missoula, Montana. *AP Photo/ Mike Albans.*

This gave Schneider pause. It didn't seem possible that vermiculite could cause occupational deaths, much less affect families at the rate that Benefield was suggesting. He returned to Seattle with three soil samples from the so-called reclamation site, had the soil analyzed, and discovered asbestos. He called the MSHA and was told that "there's nothing in the file that shows the mine was ever a problem."[58] He proceeded to call a range of federal agencies, including the CDC, EPA, OSHA, and its research arm, the National Institute of Occupational Safety and Health (NIOSH). None of these agencies confirmed that there was any connection between vermiculite and asbestos or that there had ever been an issue with the mine.

Nor had the local or state papers carried stories about widespread contamination at the Zonolite mine. Several local doctors and public health officials denied knowing about asbestos, as did the mayor of the town, who observed that surely the owner of the mine, W. R. Grace, would have warned them. That left Schneider with not much of a story—localized asbestos in the soil at the reclamation site—until he was able to get through to a regional medical center. There he discovered an unnerving reality: cases of asbestosis, mesothelioma, and asbestos-related lung cancers were taking the lives of Zonolite workers. Asbestos wasn't just in the soil: it was in the air, and it was killing people.

A return to Libby was sobering for this reporter. He met again with citizen activist Benefield, who showed him the headstones and markers at the Libby cemetery, listing a litany of people killed from asbestos. He met Carrie Detrick, diagnosed with asbestosis, who told him that she never worked at the mine, but got her asbestos death sentence from just living in the town where vermiculite was abundant.[59] Benefield then brought out files of information. She showed Schneider evidence that W. R. Grace had long known about the presence of tremolite asbestos in the vermiculite. Indeed, the company knew about asbestos when it bought the mine from its former owner, Zonolite, over thirty years earlier.

This time Schneider took more sophisticated air samples back to a lab to see if asbestos was still in the air in Libby, not just in the soil of the mine site. Continuing his search for victims, he found that, in a town of fewer than 3,000 people, at least 150 people had died from asbestos in Libby and more than 300 were sick and dying. His research uncovered that not only the company but some government agency officials had known about the asbestos contamination and done nothing for years, allowing this killer to enter the town and claim its victims. It was a chilling story—and big news.

Schneider's story broke in the *Seattle Post-Intelligencer* on November 18 and 19, 1999, with six pages of coverage and the headline "Uncivil Action: A Town Left to Die." The news story set into motion a government response to what would eventually be referred to as one of the nation's worst environmental disasters. Just as sobering as the statistics on the asbestos-related disease that permeated the town was the assertion in the story that residents who had never worked for W. R. Grace were dying of asbestosis and mesothelioma.

The news stories in the *Seattle Post-Intelligencer* not only captured public attention but prompted immediate EPA action. Schneider's series was important, not just because it drew attention to asbestos contamination, but because it pointed to the inaction of government agencies in preventing loss of life in Libby. On November 19, the second day of this big story, the headline read: "While People Are Dying, Government Agencies Pass the Buck."

The story caught the attention of Bill Yellowtail, then the Region 8 EPA administrator, who in turn contacted Steve Hawthorn, chief of the regional office's emergency response section. An on-scene coordinator for the section, Paul Peronard, was tapped to look into the validity of the story. Just four days after the articles appeared, Paul Peronard and his emergency response team arrived in Libby, Montana. EPA action was potentially warranted under the Comprehensive Environmental Response, Compensation, and Liability Act of 1980 (CERCLA, or the Superfund law). CERCLA authorizes the EPA to respond to releases of hazardous substances and to undertake both emergency removal and remedial actions. The charge to the response team was to characterize the extent of the risk to residents of Libby and to undertake response actions as appropriate.

In addition to Peronard, toxicologists, physicians, and other scientists began investigating the extent of contamination in Libby. The investigation confirmed that a large number of current and historic cases of asbestos-related disease centered in Libby, including cases of individuals who had no occupational exposure. It also confirmed the presence of significant amounts of asbestos-laced vermiculite at the former mine, the former screening plant, and the railroad loading station.

Interviews with local residents led the EPA response team to an agonizing conclusion: asbestos was most likely in Libby homes. In December 1999, the EPA collected samples of air and dust from thirty-two homes and two businesses, totaling more than six hundred samples of yards, gardens, roof insulation, and driveways.[60] The Agency for Toxic Substances and Disease Registry (ATSDR) initiated site activities in January 2000. W. R. Grace offered to pay for asbestos-health screenings and to donate $250,000 per year to the local hospital to set up a medical plan for those sickened by the mine. However, starting in March 2000, the ATSDR decided to do its own medical evaluations—wisely, as it turned out, because the results surprised everyone. Thirty percent of the 6,144 county residents who volunteered for the testing showed lung abnormalities, with 18 percent likely to have asbestos-related disease.[61] The agency would ultimately determine that Libby residents had a 40 to 60 percent higher mortality ratio for asbestos-related death than the normal population.[62]

At the same time, the Montana Department of Environmental Quality and the Department of Public Health and Human Services met with the county medical officer, local hospital officials, and local physicians. Forty years after the first diagnosis of asbestosis, it finally seemed as if all hands were on deck.

The EPA's initial response was an emergency action under CERCLA, but EPA personnel soon recognized that the hazardous conditions would require asbestos removal and long-term remediation. Under CERCLA, that meant adding Libby and W. R. Grace mining and milling sites to the

National Priorities List (NPL). Such a designation would help ensure funding and proper cleanup of asbestos-contaminated materials. The site could not be placed on the NPL, however, without the political support of the recently elected governor, Judy Martz. Although an NPL listing did not officially require the support of the state, the EPA was reluctant to move forward without it. In the previous five years, only once had the agency listed a site without a governor's approval.

A Republican and the first woman elected governor in the state of Montana, Martz faced a divided Libby constituency. On the one hand, the results of EPA and ASDTR testing revealed an unprecedented public health risk. Many community residents felt that only through the authority of the federal government's Superfund powers would the town be cleaned up. On the other hand, members of Libby's Chamber of Commerce, realtors, and business owners felt that the specter of a Superfund designation would only further depress the town's failing economic situation. The executive director of the chamber echoed the sentiments of others who felt that problems in Libby were overblown by the EPA.[63] Some residents were concerned, too, about the effect of the stigma of a Superfund designation on tourism, since Libby is located along the edge of the Kootenai National Forest.

Initially, Martz seemed sympathetic to the business voices in Libby, and to W. R. Grace. But then, in 2001, W. R. Grace's bankruptcy filing insulated the company from asbestos liability lawsuits and uncertainty increased about who would pay the cleanup costs. The Region 8 EPA office had already exceeded both the length of time and the amount of money that could be put toward an emergency removal action under CERCLA. Costs had exceeded $30 million, but as a newspaper editorial observed, that was very likely the tip of the iceberg if all contaminated homes were to be cleaned: "Maybe W. R. Grace will emerge from bankruptcy with solid finances and a newfound commitment to do right by the people of Libby. But we sure wouldn't bet the town on it. The surest way to get the job done—and done right—is through Superfund."[64]

A letter to Governor Martz from the Region 8 EPA acting administrator in December 2001 indicated that the EPA would move forward with an NPL listing even without her support: "It is EPA policy to request the concurrence of the governor of a state prior to submitting a site listing package to EPA headquarters and try to reach an agreement between EPA and Montana on an NPL listing decision. . . . A decision on whether to propose the site will then be made with a clear understanding of your position."[65] In the end, on January 16, 2002, Governor Martz not only supported the NPL designation for Libby but used the state's lone "silver bullet" under

CERCLA—the ability of a state to designate one and only one site for placement on the NPL. The silver bullet designation would fast-track cleanup efforts at a site that the governor now believed was Montana's worst environmental problem.[66]

Libby was added to the NPL in 2002, becoming the first town designated as a Superfund site. Seven years later, for the first time in the history of the agency, the EPA declared (under the Superfund law) a public health emergency in Libby in order to provide federal health care assistance for victims of asbestos-related disease.[67] By October 2013, EPA had completed removal actions at 2,000 properties and removed over one million cubic yards of contaminated soil. A year later, the EPA Remedial Investigation Report was released, documenting the nature and extent of contamination and the investigative and removal actions conducted by the EPA. In 2015, the EPA's proposed plan for the remaining portions of the Libby Asbestos Superfund site was released for public comment.[68] As of January 2016, the EPA had investigated more than 7,300 properties and completed cleanups at 2,275 of them. Contaminated soil in public places such as schools and parks was removed. The agency estimated that a few hundred more properties remained and that the final phase of the cleanup would be completed by 2020.[69] Over $600 million has been spent on cleanup operations in Libby.

THE PROSECUTION OF W. R. GRACE

On February 7, 2005, W. R. Grace and seven senior employees were indicted for knowingly exposing miners and residents to asbestos. US Attorney Bill Mercer announced the ten-count indictment alleging conspiracy, knowing endangerment, obstruction of justice, and wire fraud on the steps of the county courthouse in Missoula, Montana. Mercer referred to the asbestos exposure as a "human and environmental tragedy," and an EPA special agent called the indictment "one of the most significant criminal indictments for environmental crimes" in the agency's history.[70]

At the time of the indictment, more than 1,200 people exposed to asbestos had fallen victim to asbestos-related disease, and hundreds of miners, their family members, and Libby residents had died. The grand jury found that top executives and mine managers kept numerous studies of asbestos contamination secret and failed to warn workers and residents of the danger of ongoing asbestos exposure at the mine and in the town. W. R. Grace and Alan Stringer, a mine manager, were also indicted for trying to obstruct the EPA's investigation. Stringer faced charges amounting to seventy years in prison; senior vice president Robert Bettacchi and a former executive, Jack

FIGURE 5.1: TIMELINE OF THE ASBESTOS DISASTER IN LIBBY, MONTANA

1944: First mine inspection by the Montana Health Department identifies nuisance dust at the Zonolite plant.

1954: Zonolite expands its mill to handle over three thousand tons of vermiculite per day, making it the largest vermiculite company in the country.

1956: The industrial hygiene engineer Benjamin Wake inspects the mine and finds asbestos in the air and poor maintenance of the plant.

1959: Zonolite requests chest X-rays of all mine employees. The first official diagnosis of asbestosis is given to a miner.

1963: W. R. Grace buys the Zonolite mine and processing plants.

1964: W. R. Grace initiates annual X-ray testing of Libby employees, but does not share the results with workers.

1965: Air monitoring by W. R. Grace detects asbestos in downtown Libby.

1969: Maryland Casualty, an insurance company, sends a letter to W. R. Grace saying that not disclosing X-ray results is inhumane and violates federal law.

1980: The Comprehensive Emergency Response, Compensation, and Liability Act (the Superfund law) is passed.

1990: The Zonolite mine is closed.

1997: Les and Norita Skramstad sue W. R. Grace and win. This is the first case in which W. R. Grace is found liable.

2001: Facing over 110,000 asbestos-related lawsuits, W. R. Grace files for bankruptcy.

2005: W. R. Grace and seven top executives are indicted for endangering mine workers and residents in Libby, Montana.

2009: The EPA announces a public health emergency for Libby. W. R. Grace and executives are acquitted.

2016: W. R. Grace emerges from bankruptcy.

2017: After investigating over 7,300 properties and cleaning up 2,275 properties, the EPA issues a "last call" to residents to participate in asbestos remediation under the Superfund law.

Wolter, each faced up to fifty-five years in prison; the other four defendants faced five-year prison terms. At the same time, the company faced a fine of up to $280 million, which would amount to twice the after-tax profits from the Libby mine.[71]

In 2004, the Department of Justice obtained a trial judgment of $54 million against W. R. Grace to cover the EPA's initial cleanup activities, but the company's intervening bankruptcy filing delayed payment indefinitely.[72] On March 11, 2008, W. R. Grace agreed to pay $252.7 million—the largest cash settlement in Superfund history—to reimburse the federal government for the costs of investigation and cleanup.[73]

The criminal trial began on February 20, 2009, and concluded on May 8. The US government endeavored to prove that W. R. Grace knowingly endangered the residents of Libby, Montana. In prosecuting their case, Justice Department attorneys brought in eight longtime Libby residents who had been diagnosed with an asbestos-related disease to testify regarding their asbestos exposure pathway. Included in this testimony was a Little League coach who described piles of vermiculite on and around baseball diamonds and in dugouts, and the Lincoln County sanitarian, who noted that Libby residents commonly mixed vermiculite into their garden soil as a conditioner.

To demonstrate criminal knowing endangerment by W. R. Grace officials, government attorneys called on former employees who acknowledged that the company knew that its vermiculite contained asbestos. The government produced a host of internal W. R. Grace documents that charted the growing apprehension of the company about the lethality of vermiculite mining. Confidential memoranda revealed that as early as the 1960s the company recognized an alarmingly high rate of fatal lung diseases among the miners, but continued operations at the mine.

Testifying at length during the trial were members of the EPA's emergency response team. The on-scene coordinator, Paul Peronard, described the extensive air and other sampling done to characterize the risk in Libby. A medical officer for the EPA described exposure pathways and non-occupational asbestos-related disease in the Libby-area population. Most important to the government's case was the concept of cumulative exposures to asbestos over time increasing the risk of contracting asbestos-related disease. EPA staff testified that they believed there was an imminent health hazard to people in Libby.[74]

Alan Whitehouse, a pulmonologist from Spokane—the same doctor Andrew Schneider had interviewed ten years prior to the case—testified at trial that he had treated 1,800 Libby residents for asbestos-related disease, including 31 documented cases of mesothelioma (11 of which were from non-occupational exposure). This testimony was significant, because mesothelioma, as described earlier, rarely occurs without exposure to asbestos.[75]

The lead counsel for W. R. Grace, David Bernick, emphasized that the company did not conspire to hide asbestos problems in Libby and that those dangers were widely known by everyone, including the state of Montana and the EPA. Prosecutors relied on internal company memos to suggest an intentional coverup by W. R. Grace, while the defense team for the company argued that the paper trail showed the efforts of a company trying to make the workplace safer. The jury in the US District Court in Missoula, Montana, deliberated just two days before unanimously concluding that the asbestos contamination was not a criminal act.

Different theories have been offered regarding the verdict. By some accounts, government lawyers did not handle the case well; at one point, for instance, they were chastised by US District Judge Donald Malloy for not understanding the evidence they were presenting.[76] Some internal documents from W. R. Grace were ruled inadmissible by the judge, including one memo that calculated the costs of asbestos deaths in Libby. However, the timeline for the case was also a problem. The knowing endangerment criminal provisions of the Clean Air Act did not take effect until 1990, the same year that W. R. Grace closed the mine. Also, the statute of limitations required that the government prove that laws were violated after 1999, or within five years of the 2005 indictment. Additionally, the testimony of the government's star witness, Robert H. Locke, a former W. R. Grace executive who testified that the company actively worked to hide their knowledge about asbestos deaths, was denounced by Judge Malloy. Calling the government's failure to disclose the number of meetings that prosecutors had with Locke an "inexcusable dereliction of duty," the judge instructed the court to use "great skepticism" in evaluating Mr. Locke's testimony.[77]

The outcome of the case also rested in part on the instructions to the jury and on the court's interpretation of the knowing endangerment violations under the Clean Air Act. According to an analysis offered by Kris McClean, the lead prosecutor in the case, and two other environmental attorneys, the judge's instructions tipped the outcome of the case in favor of W. R. Grace. The following instruction was given to the jurors:

> For a defendant to be found guilty of the crime of knowingly releasing or willfully causing the release of a hazardous air pollutant into the ambient air, namely, asbestos and at the time, knowingly placing another person in imminent danger of death or serious bodily injury

the government must prove each of the following elements beyond a reasonable doubt:

First, that the defendant under your consideration knowingly released or willfully caused the release into the ambient air a hazardous air pollutant, namely asbestos;

Second, that the release occurred after November 3, 1999;

Third, that the defendant under your consideration knew that by knowingly releasing or willfully causing to be released a hazardous air pollutant, namely asbestos, the defendant placed another person in imminent danger of death or serious bodily injury.

To find a defendant guilty of the charges [knowing endangerment] all of you must unanimously agree as to the same specific release or releases, occurring for the first time after November 3, 1999, that placed another person or other persons in imminent danger of death or serious bodily injury.[78]

These instructions forced jurors to identify, with unanimous agreement and beyond a reasonable doubt, that particular releases had placed a person in a situation that would more likely than not cause serious bodily injury, and that those releases occurred after November 3, 1999—the date that the EPA began its investigations. Lawyers for the W. R. Grace defendants were quick to highlight this standard in their closing arguments, pointing out that the release had to be specific, identifiable, directed to an individual person, and more likely than not to cause death or serious bodily injury. Given these instructions, the jury had little choice but to acquit W. R. Grace defendants of knowing endangerment under the Clean Air Act.[79] By the time of the trial, only four defendants out of the seven in the original indictment were left. Alan Stringer, the general manager of the Zonolite mine, died of cancer in 2007.

Despite winning in court, W. R. Grace continues to pay for its toxic legacy. In 2014, the company paid over $63 million to the US government under its bankruptcy plan of reorganization—a sum that was meant to resolve claims for environmental cleanups at thirty-nine sites in twenty-one states. The company will continue to be responsible for all of the sites it owns or operates, as well as those that were not part of this agreement. This sum is in addition to the $252 million paid to the EPA for the cleanup of asbestos contamination in Libby.[80] As Robert G. Dreher, acting assistant attorney general, noted, "The Justice Department is committed to holding polluters responsible for their environmental legacy, and won't just walk away leaving taxpayers to pick up the tab."[81]

W. R. Grace filed for bankruptcy due to asbestos-related claims on April 2, 2001.[82] It would take seven years before the company agreed to create an asbestos trust to pay settlements for the more than 129,000 cases already filed, as well as new cases.[83] On February 3, 2014, W. R. Grace emerged from Chapter 11 bankruptcy, after establishing two trusts worth $4 billion to compensate victims and property owners exposed to asbestos.[84] Staying in bankruptcy had provided W. R. Grace with a reprieve from having to defend itself against thousands of asbestos claims, while at the same time nearly doubling its business from $1.8 billion to $3.2 billion in revenues and increasing its stock price from less than $2 a share to more than $90. As the lead asbestos lawyer in the bankruptcy noted, "W. R. Grace has been very successful commercially, in part because it has not had to spend its energy and money litigating asbestos issues in the tort system for the last 12 years."[85] In financial terms, the company appears to have prospered even in litigating asbestos claims.

CONCLUSION

The story of Libby, Montana, has heroes and villains aplenty. Though W. R. Grace successfully defended itself against the charge of knowingly endangering the lives of its employees and the town residents in a court of law, the court of public opinion will forever see the company as acting with a callous disregard for the health of those who had the misfortune to breathe the dust from its mine. Internal company documents clearly illustrate that W. R. Grace and its predecessor, Zonolite, were aware that the dust was more than just a nuisance—they knew that it was deadly.

For at least thirty years, both companies did little to protect workers and instead sought to exploit the economic potential of its asbestos-laden vermiculite. But worse, W. R. Grace made a conscious decision to keep its knowledge of the composition of the deadly dust from employees, people at other processing plants across the country, and purchasers of its products. Even W. R. Grace's insurance company called the company out on its behavior, which it branded as inhumane and illegal. Every person in the chain of command who had the opportunity to inform employees of the hazards, to stop the dust from spreading to the homes of workers, to keep piles of vermiculite out of the Libby community, and to prevent the spread of asbestos contamination to other processing facilities and to the homes of anyone using Zonolite or W. R. Grace products, will have to live with their decisions.

Given the hazards of occupational exposure to asbestos, the widespread use of asbestos in homes and buildings suggests that many more Americans have experienced environmental exposure to asbestos. Although asbestos consumption in the United States peaked in 1973 and dropped dramatically in the last three decades, many homes and buildings still contain asbestos, Zonolite insulation, and asbestos-containing materials.[86]

Heroes are found, too, in this story. Prominent among them are Gayla Benefield, Andrew Schneider, and Les Skramstad. Benefield was determined to bring the failures of W. R. Grace to the public's attention. Working tirelessly on behalf of not only her parents (both of whom died) but also herself, her husband, and the hundreds of others suffering from asbestos-related diseases, she was a tireless advocate for getting Libby cleaned up and holding W. R. Grace responsible. She had file drawers' worth of information about the company, which she shared with Schneider as he developed his story. When the EPA finally came to Libby, Benefield was the one who championed the efforts of the agency as well as the listing of the town as a Superfund site. She won her own lawsuit and attended the company's criminal trial. In an interview with *Democracy Now,* Benefield commented:

> Bluntly, in reading the [W. R. Grace internal] documents, Libby was collateral damage. The company was here to make money, and the men who worked here were simply collateral damage. And it was easier to pay a small amount of reimbursement to families or to men who became ill than to simply shut down the mine. So, for years, it was a cover-up. And the upper management, the men that were on trial, were all well aware of the dangers posed.[87]

Andrew Schneider's role cannot be disregarded. His investigative reporting was the triggering event that brought Libby before the public eye and finally forced government agencies to act. Without his determined efforts, Libby's plight might never have been brought to light. Perhaps as important as his series of articles was the way in which he wrote about W. R. Grace and government inaction. He masterfully juxtaposed the shortcomings, even arrogance, of the company against the innocence of the workers, their families, and the affected communities. The communications scholar Steven Schwarze would later observe of these news articles:

> W. R. Grace's villainy comes not only from lying about specific facts, but in maintaining that lie over time, perhaps over the stretch of multiple managers such that W. R. Grace the company is to blame, not a particular person. Finally, this juxtaposition heightens the innocence of the victim, who did not know he was sick and yet was knowingly

exposed to "all that poison." . . . The juxtaposition of the company's words and actions with the ultimate effects on human health work rhetorically to heighten outrage toward W. R. Grace and emphasize the innocence of the victim.[88]

Les Skramstad, diagnosed with asbestosis in 1996, became the first worker to sue W. R. Grace in a jury trial and win. He, with Benefield, became an advocate for the town and a constant reminder of the company's failure to warn its employees about the dangers of the "nuisance" dust. Skramstad later discovered that his wife, Norita, and three of his four children also had the disease, a heavy burden to bear. He said, "I loved my job at Zonolite. . . . Even if they told me it was endangering my life, I would probably have stayed. But if they told me it was killing my wife and my family, too, I would have run like hell."[89] Les Skramstad died before the oft-delayed criminal trial against W. R. Grace and its top officials. His wife, Norita, brought his hat to the trial, in accordance with his wishes.[90]

DISCUSSION QUESTIONS

1. What similarities and differences do you see between the stories of the Zonolite mine in Libby, Montana, and the Upper Big Branch mine in West Virginia (Chapter 4)?
2. In what ways did W. R. Grace fail its workers and the town of Libby?
3. Using the characteristics of environmental heroes and villains described in Chapter 1, would you identify anyone in this story as a villain or a hero?
4. Does the slow evolution of a disaster, such as the one in Libby, make it harder to gain public or governmental attention? Why or why not?

NOTES

1. US Environmental Protection Agency (EPA), Region 8, "Background on the Libby Asbestos Site," March 24, 2015, www2.epa.gov/region8/background-libby-asbestos-site (accessed June 4, 2015).

2. Andrew Schneider and David McCumber, *An Air That Kills: How the Asbestos Poisoning of Libby, Montana, Uncovered a National Scandal* (New York: G. P. Putman, 2004), 2.

3. EPA, "US Federal Bans on Asbestos," last updated December 19, 2016, www.epa.gov/asbestos/us-federal-bans-asbestos (accessed May 29, 2015).

4. The Mesothelioma Center, "Types of Asbestos," Asbestos.com, last modified August 28, 2017, www.asbestos.com/asbestos/types.php (accessed April 20, 2014).

5. The EPA refers to the asbestos found in Libby, Montana, as tremolite-actinolite series asbestos, or Libby amphibole asbestos (LA). See EPA, "Superfund Site: Libby Asbestos Site, Libby, MT," https://cumulis.epa.gov/supercpad/Cursites/csitinfo.cfm ?id=0801744&msspp=med (accessed June 19, 2017).

6. US Department of Labor, Occupational Safety and Health Administration (OSHA), "Substance Technical Information for Asbestos—Non-Mandatory," www .osha.gov/pls/oshaweb/owadisp.show_document?p_table=STANDARDS&p_id =10870 (accessed June 19, 2017).

7. Pascale Krumm, "The Health Effects of Asbestos," *Journal of Environmental Health* 65, no. 2 (September 2002): 46.

8. EPA, *Asbestos Fact Book,* A-107/86-002 (Washington, DC: EPA, Office of Public Affairs, May 1986), 3.

9. The Mesothelioma Center, "Asbestosis," Asbestos.com, last modified March 31, 2017, www.asbestos.com/asbestosis/ (accessed June 12, 2015).

10. Krumm, "The Health Effects of Asbestos."

11. Center for Asbestos Related Disease, "Smoking and Asbestos-Related Disease: Facts," www.libbyasbestos.org/education/patient-education/health-risks/smoking/.

12. American Cancer Society, "Asbestos and Cancer Risk: What Is Asbestos?" updated September 15, 2015, www.cancer.org/cancer/cancercauses/othercarcinogens /intheworkplace/asbestos (accessed May 29, 2015).

13. American Cancer Society, "Survival Statistics for Mesothelioma," updated February 17, 2016, www.cancer.org/cancer/malignantmesothelioma/detailedguide /malignant-mesothelioma-survival-statistics (accessed June 12, 2015).

14. American Cancer Society, "What Are the Risk Factors for Malignant Mesothelioma?" updated February 17, 2016, www.cancer.org/cancer/malignantmesothelioma /detailedguide/malignant-mesothelioma-risk-factors (accessed June 12, 2015).

15. American Cancer Society, "Asbestos and Cancer Risk."

16. Johns Manville Corporation, "History and Heritage: Company History," 2016, www.jm.com/en/our-company/history-heritage-berkshire-hathaway/company -history/.

17. Ibid.

18. Johns-Manville Company, ad in *Saturday Evening Post* 189, no. 3 (February 24, 1917): 56.

19. Barry Castleman, "The Question of Asbestos in the United States of America," paper presented at the Fiocruz Conference on Asbestos and Asbestos Substitutes, Rio de Janeiro, September 3, 1998, http://ibasecretariat.org/bc_asbestos_usa .php (accessed May 29, 2015).

20. Alda McKnight, "Asbestos in America—From World War II to Now," American History USA, January 25, 2013, www.americanhistoryusa.com/asbestos-in -america-from-world-war-ii-to-now/ (accessed June 19, 2017).

21. Bill Burke, "Shipbuilding's Deadly Legacy: A Special Report," *Virginian Pilot,* May 6, 2001. Excerpts of the original story are available at The Mesothelioma Center, www.mesotheliomacenter.org/mesothelioma-news/2001/05/09/shipbuildings -deadly-legacydisease-9000-sick-or-dead/ (accessed April 29, 2014).

22. Ibid.; also quoted in Denise Scheberle, *Federalism and Environmental Policy: Trust and the Politics of Implementation* (Washington, DC: Georgetown University Press, 2004).

23. William Nicholson, "Occupational Exposure to Asbestos: Population at Risk and Projected Mortality 1980–2030," *American Journal of Industrial Medicine* 3 (1982): 259–311, cited in Stephen Carroll, Deborah Hensler, Allan Abrahamse, Jennifer Gross, Michell White, Scott Ashwood, and Elizabeth Sloss, *Asbestos Litigation Costs and Compensation: An Interim Report* (Santa Monica, CA: RAND, 2002), www.rand.org/publications/DB/DB397/ (accessed March 28, 2003).

24. Burke, "Shipbuilding's Deadly Legacy."

25. Environmental Working Group, "The Asbestos Epidemic in America," section 1 in "Understanding Asbestos," www.ewg.org/asbestos/facts/fact1.php (accessed June 5, 2015).

26. Jacek M. Mazurek, Girija Syamlal, John M. Wood, Scott A. Hendricks, and Ainsley Weston, "Malignant Mesothelioma Mortality—United States, 1999–2015," *Morbidity and Mortality Weekly Report* 66 (2017): 214–218, Centers for Disease Control and Prevention, doi:http://dx.doi.org/10.15585/mmwr.mm6608a3.

27. The Mesothelioma Center, "Johns Manville," Asbestos.com, last modified April 28, 2017, www.asbestos.com/companies/johns-manville.php (accessed February 9, 2015).

28. Mesothelioma and Asbestos Awareness Center, "A Brief History of Asbestos Litigation," July 27, 2016, www.maacenter.org/blog/community/a-brief-history-of-asbestos-litigation/ (accessed June 20, 2017).

29. The Mesothelioma Center, "Mesothelioma and Asbestos Trusts," Asbestos.com, last modified August 27, 2017, www.asbestos.com/legislation/trust-fund.php (accessed May 29, 2015).

30. Libby Legacy Project, "Libby Timeline," November 27, 2012, www.libbyschools.org/sites/default/files/page-files/LLP_112612_Condensed.pdf (accessed June 12, 2015).

31. Schneider and McCumber, *An Air That Kills*, 40.

32. Ibid., 45.

33. Libby Legacy Project, "Libby Timeline."

34. Ibid.

35. Andrea Peacock, *Wasting Libby: The True Story of How the W. R. Grace Corporation Left a Montana Town to Die (and Got Away with It)* (Petrolia, CA: AK Press/CounterPunch, 2010), 83.

36. Quoted in ibid., 109.

37. Libby Legacy Project, "Libby Timeline."

38. Quoted in Peacock, *Wasting Libby*, 42.

39. Ibid., 108.

40. Libby Legacy Project, "Libby Timeline."

41. Quoted in ibid.

42. Andrea Peacock, *Libby, Montana: Asbestos and the Deadly Silence of an American Corporation* (Boulder, CO: Johnson Books, 2003), 86.

43. Ibid., 85.

44. Andrea Barnett and Maryanne Vollers, "Libby's Deadly Grace," *Mother Jones* 25, no. 3 (May/June 2000), www.motherjones.com/environment/2000/05/libbys -deadly-grace/.

45. Peacock, *Wasting Libby*, 87.

46. Quoted in Libby Legacy Project, "Libby Timeline."

47. Peacock, *Wasting Libby*, 40.

48. Quoted in ibid., 140.

49. Peacock, *Libby, Montana*, 133.

50. *Orr v. State of Montana*, Supreme Court of Montana, 02-693, December 14, 2004.

51. Ibid., at 37.

52. Tristan Scott, "Libby Asbestos Victims Reach $25 Million Settlement with State," *Flathead Beacon*, January 19, 2017.

53. The General Mining Law of 1872 is still the law governing mining on federal public lands. It allows companies to buy permits from the government to extract gold, silver, and other precious metals. Under this law, companies can obtain mining permits for very little money, pay no royalties, and leave the land and water resources contaminated and unreclaimed. It's an environmental story well worth telling, but beyond the scope of this chapter.

54. Schneider and McCumber, *An Air That Kills*, 10.

55. Ibid., 11.

56. Ibid., 21.

57. Ibid., 13.

58. Ibid., 15.

59. Ibid., 23.

60. Bill Yellowtail, "Asbestos Contamination in Libby, Montana," testimony before the US Senate Committee on Environment and Public Works (field hearing), February 16, 2000, Senate Hearing 106-950.

61. Peacock, *Wasting Libby*, 223.

62. US Department of Justice, "US v. W. R. Grace & Co., Environmental and Public Health Disaster at Libby, Montana," May 14, 2015, www.justice.gov/enrd/us-v-w -r-grace-co (accessed September 22, 2017).

63. Kathleen McLaughlin, "Martz Wants Consensus in Divided Libby," *Missoulian*, June 22, 2001.

64. "Libby Needs Certainty of Cleanup" (editorial), *Missoulian*, June 14, 2001.

65. Kathleen McLaughlin, "EPA Asks Martz to Back Libby Superfund Designation," *Helena Independent Record*, December 14, 2001.

66. Kathleen McLaughlin, "Martz Makes Silver Bullet Status Official," *Missoulian*, January 16, 2002.

67. EPA, "Libby Public Health Emergency," June 17, 2009, www2.epa.gov/region8 /libby-public-health-emergency (accessed April 29, 2014).

68. Region 8 EPA, "Libby Asbestos: What's New," June 11, 2015, www2.epa.gov /region8/libby-asbestos (accessed June 17, 2015).

69. EPA, "Superfund Site: Libby Asbestos Site Libby, MT," 2017, https://cumulis
.epa.gov/supercpad/Cursites/csitinfo.cfm?id=0801744&msspp=med (accessed June
21, 2017).

70. Andrew Schneider, "W. R. Grace Indicted in Libby Asbestos Deaths," *St.
Louis Post-Dispatch,* February 7, 2005.

71. Associated Press, "Charges Issued over Asbestos at a Mine," *New York Times,*
February 5, 2005.

72. US Department of Justice. "US v. W. R. Grace & Co., Environmental and Public Health Disaster at Libby, Montana."

73. Region 8 EPA, "Libby Major Milestones," www2.epa.gov/region8/libby-major
-milestones (accessed April 29, 2014).

74. Linda Kato, Kris A. McLean, and Eric Nelson, "Toxic Torts, Knowing Endangerment, and the W. R. Grace Prosecution," *Environmental Crimes—2012,* 60, no.
4 (2012): 107–124.

75. Ibid., 121.

76. Kirk Johnson, "Chemical Company Is Acquitted in Asbestos Case," *New York
Times,* May 8, 2009.

77. Ibid.

78. Kato, McLean, and Nelson, "Toxic Torts, Knowing Endangerment, and the
W. R. Grace Prosecution," 121–122.

79. Ibid., 124.

80. US Department of Justice, "W. R. Grace Pays over $63 Million Toward the
Cleanup and Restoration of Hazardous Waste Sites in Communities Across the
Country" (press release), February 5, 2014, www.justice.gov/opa/pr/wr-grace-pays
-over-63-million-toward-cleanup-and-restoration-hazardous-waste-sites (accessed
September 22, 2017).

81. Ibid.

82. W. R. Grace, "Chapter 11 and Asbestos," https://graceW.R. Grace.com/en-us/
Pages/chapter-11-and-asbestos.aspx (accessed June 5, 2015).

83. Andrew Schneider, "W. R. Grace Readies $3 Billion for Asbestos Victims,"
Seattle Post-Intelligencer, April 7, 2008.

84. Catherine Ho, "W. R. Grace Emerges from Chapter 11 Bankruptcy After More
Than 12 Years," *Washington Post,* February 4, 2014.

85. Quoted in Peg Brickley, "Five Takeaways from the W. R. Grace Bankruptcy,"
Wall Street Journal, February 3, 2014.

86. James Alleman and Brooke Mossman, "Asbestos Revisited," *Scientific American* (July 1997): 70–75.

87. "W. R. Grace Acquitted in Libby, Montana Asbestos Case" (transcript of interview), *Democracy Now,* May 12, 2009, www.democracynow.org/2009/5/12/wr
_grace_acquitted_in_libby_montana.

88. Steven J. Schwarze, "Juxtaposition in Environmental Health Rhetoric: Exposing Asbestos Contamination in Libby, Montana," Communication Studies

Faculty Publications Paper 13 (Missoula: University of Montana, Summer 2003), http://scholarworks.umt.edu/communications_pubs/13 (accessed June 16, 2015).

89. Schneider and McCumber, *An Air That Kills,* 3.

90. Andrew Schneider, "Criminal Trial Begins in Montana Asbestos Case," *Seattle Post-Intelligencer,* February 23, 2009.

CHAPTER 6

Environmental Heroes

NELSON, RUCKELSHAUS, AND BONDS

Unlike the stories in the previous four chapters, not all environmental stories end in disaster. Many stories are hopeful, with characters who prevail in keeping workplaces safe and environments protected. The focus of this chapter is on what individuals have accomplished. Offering a transition between stories of disasters and stories of optimism and courage, the chapter highlights three stories of impressive accomplishments by three environmental heroes: Gaylord Nelson, William Ruckelshaus, and Judy Bonds.

I chose to tell the stories of these individuals because each of them shaped our environmental heritage, laws, policies, or agencies in profound ways. Each, as we will learn, exhibited characteristics of an environmental hero, and each faced challenging circumstances in advancing environmental goals. Two of them, Gaylord Nelson and William Ruckelshaus, made their mark on environmental history during the first environmental decade of the 1970s, while Bonds was active around the turn of this century. Nelson represents extraordinary vision and dedication, Ruckelshaus represents organizational skill and integrity, and Bonds represents courage and initiative. Their stories provide a sample of what individuals, acting on their own or inside of organizations, can do to galvanize attention to an environmental issue and fight the good fight with integrity, grit, and wisdom. These three remarkable individuals changed the course of environmental protection, but they are far from the only environmental heroes of the last fifty years—the stories of many other environmental heroes could fill the pages of this book if space and time permitted.

The chapter begins with a quick overview of the characteristics of environmental villains and heroes, as presented in Chapter 1, and then tells the stories of each of these heroes. In my concluding observations about the common elements of each story, I hope to establish a framework for advancing environmental protection that the rest of us can use to become involved.

REVISITING THE CHARACTERISTICS OF ENVIRONMENTAL VILLAINS AND HEROES

Environmental Villains

Whether an individual is an environmental villain or hero is in the eye of the beholder, so it is ultimately up to each of us to assess the villainy that may appear in each of these stories. To that end, it's worthwhile to briefly revisit the characteristics of villainy described in Chapter 1 in light of the stories told in Chapters 2 to 4 (see Box 1.1).

Villainous behavior is seen in continued disregard for the environment and the discounting of human health and safety risks. Villains are powerful individuals who accept unreasonable risks in running their operations, understand the potential devastation if safety protocols are not followed, and yet do not deviate from their dangerous practices. They obfuscate the truth, hiding safety problems or environmental dangers from both workers and residents in nearby communities. They avoid complying with regulatory requirements, even to the point of covering up safety or public health violations. In short, they persist in doing little or nothing to protect workers or the environment. If they are corporate or organizational leaders, environmental villains often seek to shift blame away from themselves after disaster strikes, claiming that they had no control over what was happening, and that the blame rests with the workers or lower-level managers.

For example, while he was CEO of Massey Energy, Don Blankenship often disregarded miner safety, as was evident in many of his actions (see Chapter 4). Perhaps the best illustration of his disregard was his practice of keeping two separate books when safety issues were identified—one for mine managers, another for federal and state inspectors. Similarly, the plant managers at the W. R. Grace facility in Libby, Montana, looked the other way as their workers breathed lethal doses of asbestos and let them incorrectly assume that the dust was just a nuisance (Chapter 5). One plant manager, Earl Lovick, knew for decades that asbestos was in the mine, yet he was willing to permit his workers, their families, and the community to be exposed. When the Union Carbide pesticides facility in Bhopal, India, sent deadly gas into the air, killing thousands of people, the Indian government sought the extradition of Union Carbide's CEO at the time, Warren Anderson, to be tried on charges of culpable homicide. Advocates for the victims charged that Anderson was well aware of the dangerous conditions in Bhopal (see Chapter 2). Not only did Anderson avoid charges, but it would be twenty-six years before the Union Carbide officials in charge of the Bhopal plant were prosecuted and found guilty. Likewise, Tony Hayward, BP's CEO at the time of the BP oil spill, had overseen an organization with a long history of violations of environmental and safety regulations—more than any other oil company operating in the United States.

In sum, these stories offer food for thought about individual behavior and the potential for organizations and the people who run them to move to the dark side.

Environmental Heroes

Heroes exhibit moral fortitude, acting with integrity in the face of pressure to do otherwise (see Box 1.3). Thoughtful and analytical, they bring wisdom to their actions and employ facts and the best available science to present their case. Environmental heroes are more than role models (though they are certainly that as well) in that they exhibit courage in the face of seemingly insurmountable odds. They fight the good fight, perhaps for years, and remain dedicated to what they believe is right, even in the face of threats, including physical dangers, the vitriol of others in the community or organization, the loss of their livelihood, or other economic, physical, or psychological costs. This persistence is combined with resourcefulness as they thoroughly and systematically assess what is right, what needs to be done, and how to proceed, often in innovative and creative ways.

Although villains loomed large in the previous four environmental stories, heroes played a part in those stories as well. For example, Gayla Benefield, Les Skramstad, and others fought tenaciously to expose W. R. Grace's unsafe practices in Libby, Montana (Chapter 5). They did not give up, even though they were pitted against the town's major employer, a host of its attorneys, an uninterested media, and do-little state and federal regulatory agencies. Without Benefield's persistence and determination, it is unlikely that Andrew Schneider from the *Seattle Post-Intelligencer* would have pursued Libby's story. Paul Peronard, the Region 8 EPA emergency response coordinator—who himself was highly regarded for his on-site efforts at Libby—would later credit Benefield and other Libby residents for successful community coordination efforts.

Or recall Raajkumar Keswani, writing for the *Rapat Weekly*, a small paper in Bhopal, who warned about the dangers at the Union Carbide plant (see Chapter 2). Two years before the disaster, Keswani published his first article, titled "Save, Please Save This City." He repeated warnings about inadequate safety standards in follow-up articles in multiple publications, quoting extensively from Union Carbide's own report documenting problems at the plant.[1] When no changes to plant operations were forthcoming, Keswani wrote the chief minister of the state of Madhya Pradesh, Arjun Singh, declaring his intention to continue to warn people about the dangers at the plant. Showing his resolve, he wrote: "I will not give up, I will fight with firm determination—I will not let this city turn into Hitler's gas chamber."[2]

As remarkable as these heroes were, it is useful to explore other heroes whose impact was far-reaching. The following sections tell the stories of three such heroes, beginning with Senator Gaylord Nelson.

GAYLORD NELSON: FOUNDER OF EARTH DAY

The first story is of Gaylord Nelson, an ardent environmental leader who served as a US senator from Wisconsin and also as governor of that state. Motivated by his Wisconsin country upbringing, and then by an oil spill in Santa Barbara in 1969, Nelson founded Earth Day and was the architect of some of the first national environmental laws. However, his environmental accomplishments were not gained easily in the 1960s and 1970s. Indeed, what Nelson achieved would not have come about without his determination, persistence, and willingness to be scorned by those who thought his environmental agenda was radical and divorced from the values of Americans. Nelson did not falter in his belief that enacting tough new national laws protecting the environment was possible and, more important, that ordinary people could be persuaded to have a voice in shaping environmental outcomes.

Early Years: A Push for Environmental Issues

Nelson spent much of his public life advocating for environmental protection. As a small-town boy born in 1916 in Clear Lake, Wisconsin, he grew up with an appreciation for the richness of nature. After serving ten years in the Wisconsin state legislature, he was elected governor of the state in 1958.

As governor, he pushed for the state to preserve open space, expand outdoor recreational opportunities, and protect wildlife habitat. He persuaded the legislature to pass a one-cent tax per package of cigarettes and to use the revenue for the state's Outdoor Recreation Acquisition Program, a $50 million state land preservation program. This was the first conservation program of its kind in the nation.[3]

Elected to the US Senate in 1962, Nelson gained an appointment to the Senate Interior and Insular Affairs Committee, where he could pursue his environmental agenda. As a freshman senator, Nelson pushed for congressional and executive action on environmental protection. Aware of the bully pulpit that a president commands, he wrote to President John F. Kennedy to encourage him to undertake a national tour highlighting conservation as a national priority. In his letter, Nelson urged the president to tell the public that "there is no domestic issue more important to America in the long run than the conservation and proper use of our natural resources, including fresh water, clean air, tillable soil, forests, wilderness, and habitat for wildlife, minerals and recreational assets."[4]

In pressing for this unprecedented presidential conservation tour, Nelson argued that the failure to protect the environment had in fact been a failure of political leadership on environmental issues. He saw Kennedy as a champion for conservationists. Through a series of speeches throughout the country, the president would shine a light on the plight of America's natural resources. Nelson believed that Kennedy could "shake people, organizations, and legislators" and that this public outcry would, in turn, galvanize Congress to acknowledge environmental issues.[5] In September 1963, Kennedy visited eleven states to promote conservation. The tour did little, however, to stimulate either public or political interest in the issue.[6] Crowds were small, and the press was more interested in asking about foreign affairs—particularly a nuclear test-ban treaty with the Soviet Union—than about conservation. Even if Nelson had been able to persuade Kennedy to undertake another conservation tour, it was sadly not to be. The president was assassinated just two months later.

Knowing that his influence as a newly elected senator was limited, the undaunted Nelson nonetheless pressed his environmental concerns in the US Senate. During a speech on the Senate floor, Nelson framed the state of the country's environment as a national issue of major importance, saying, "We need a comprehensive and nationwide program to save the national resources of America. Our soil, our water, and our air are becoming more polluted every day. Our most priceless natural resources—trees, lakes, rivers, wildlife habitats, scenic landscapes—are being destroyed."[7] In 1964, Nelson was part of a group of legislators who sponsored the Wilderness Act, which safeguarded millions of acres of federal land. Later he worked to pass the Wild and Scenic Rivers Act of 1968. But in Nelson's view, much more needed to be done to protect the quality of the country's water, air, and land.

Passing the Wilderness Act in 1964 and the Wild and Scenic Rivers Act in 1968 was a legislative accomplishment. However, convincing Congress to protect natural resources was an easier "sell" than establishing stringent national laws to control pollution from power plants, sewage treatment facilities, and large industries. Regulating the end-of-pipe pollution generated by the captains of industry was an entirely different matter, as it represented a direct challenge to the way America's economy functioned. In the 1960s, impassioned Senate speeches from Nelson and like-minded colleagues—most notably Edmund Muskie (D-ME), who himself was an environmental champion—had failed to push this thorny issue onto the mainstream political agenda. Realizing that postwar America was pushing environmental limits, Nelson observed that the country was facing the "darkening cloud of environmental pollution," and that the "mindless pursuit of quantity is destroying, not enhancing, the opportunity to achieve quality in our lives."[8] In Nelson's opinion, more was needed to advance the issue—such as a groundswell of public opinion demanding that environmental deterioration be stopped.

The Idea for Earth Day Is Born

Nelson realized that creating a groundswell of public support for environmental protection would not be easy. Then a focusing event became a catalyst for Nelson: the Santa Barbara oil spill. The year was 1969, six years after President Kennedy's conservation tour, which Nelson had hoped would "wake up" Washington and the American public. Nelson traveled to California to inspect what at the time was the largest oil spill in the nation's history. In an eerie precursor to the BP oil spill, a blowout had erupted below the drilling platform of the Unocal Corporation and would ultimately spew more than three million gallons of crude oil before it was plugged.[9] Though there were no cameras at the ocean floor, as there would be in the blowout of the Macondo well, images of oil-soaked birds dying on the shores, waves thick with oil, and beaches coated with thick sludge greeted American television viewers and made their way into national newspapers and news magazines.

As he stood on the California shoreline inspecting the effects of the oil spill, Nelson realized that this tragedy had provided the impetus to act—now was the time to bring public concern over the spill to the attention of what he perceived as a lethargic political community.[10] It would be people, he reasoned, not the president, who would force pollution control onto the congressional agenda and usher in a host of new environmental laws. The Santa Barbara oil spill, like the disasters described in previous chapters, would become a triggering event that prompted environmental issues to move onto national and state policymaking agendas.

However, he needed a way to demonstrate to Congress that people across the country wanted action on environmental issues. Luckily, one grassroots model was making national headlines at the time: the teach-ins about the Vietnam War being held on college campuses. Nelson envisioned using similar teach-ins for environmental causes. He returned to Washington, DC, and developed the concept of Earth Day, then announced the event during a speech in Seattle on September 9, 1969. The news services picked up the story, and soon articles appeared across the country.

By November 1969, a date had been chosen: April 22, 1970. The organizers, hoping to get college students involved, picked a date that would come before the pressure of final exams at the end of a typical spring semester. But enthusiasm for the event went well beyond college campuses. Elementary and high school students wrote to Nelson, as did churches and community groups. The response was overwhelming. Nelson and his Senate office staff were inundated with requests for information. It seemed that everyone wanted to participate in a teach-in for environmental protection.

In a show of bipartisan support for Earth Day, Nelson asked Congressman Paul McCloskey (R-CA) to serve on a steering committee to determine how to handle the large surge of public interest in holding an environmental event. When the group recommended creating an organization to assist with requests, Nelson appointed himself cochairman (with McCloskey) of a newly formed nonprofit organization, Environmental Teach-In. The organization served as a resource for organizers of local events. Denis Hayes, a twenty-five-year-old student from Harvard, became national coordinator. In the span of just seven months, Nelson and his team created something that would forever serve as a model for civic environmentalism. From the beginning, Nelson saw Earth Day as a grassroots movement that would sponsor local events reflecting the interests of participants and become a political force that could not be ignored by complacent legislators. He saw Earth Day as more than just a one-day event; he hoped it would be the start of a movement that would pressure governments to address environmental issues.

Earth Day, April 22, 1970, sparkled with the energy of people coming together across the country. Nelson's vision had been genius: events were staged by grassroots organizations in dozens of US communities. From the beginning, Nelson insisted that the day was not to be a uniform national protest but would be devoted to local, "old-fashion political action." He would later recount, "I took a gamble, but it worked."[11] Earth Day not only exceeded his expectations, it was a huge success. Teach-ins and marches throughout the country were held on an estimated two thousand college campuses, in thousands of communities, and in ten thousand elementary and secondary schools. An estimated twenty million people participated in the first Earth Day—an astounding number by any standard, representing almost 10 percent of the US population at the time.[12] In comparison, the March for Science, held forty-seven years later on Earth Day 2017, had an estimated 1.3 million participants in six hundred locations—still successful, but representing less than 1 percent of the population.[13]

Thousands of people marched in New York City, and for a few hours a small part of the city was off-limits to motor vehicles as participants demonstrated for a cleaner environment. Similar marches were held in Chicago, Atlanta, Miami, Denver, and Philadelphia—just to name a few. True to its organizational name (Environmental Teach-In), audiences listened to scientists, ecologists, politicians, and teachers warn that now was the time to take responsibility for reducing pollution and protecting the environment. Though students led many events, people of all ages and political affiliations could be seen in the crowds, including politicians as different in their political views at the time as Senator Edward Kennedy (D-MA) and Senator Barry Goldwater (R-AZ).

Nelson, who had worked tirelessly to create this single day, spoke at eighteen events across the country.[14] In Denver, he emphasized that environmental protection should be inclusive, embrace the concept of sustainability, and respect social equity concerns. "Our goal is not just an environment of clean air and water and scenic beauty. The objective is an environment of decency, quality and mutual respect for all other human beings and living creatures."[15]

The peaceful and inclusive nature of the first Earth Day stood in contrast to the tumult that frequently accompanied political protests in the late 1960s and early 1970s, in particular the antiwar protests that swept the country. On April 30, 1970, just a week after the first Earth Day, President Richard Nixon announced on national television that he had ordered the invasion of Cambodia, widening the already unpopular Vietnam conflict. This prompted heated protests across college campuses. Perhaps the most infamous of these antiwar demonstrations was at Kent State University, where four students were killed and nine others wounded on May 4, prompting student strikes at more than 450 colleges and universities.

Against this backdrop of political upheaval over the Vietnam War, the success of Earth Day and its political sway over the US Congress are even more remarkable. The event was one of those rare historical moments when Republicans and Democrats, rural and urban residents, young and old people, aligned to signal that the environment was worth protecting. As Nelson would observe, "The objective of Earth Day was fully accomplished. My objective was to have a massive nationwide demonstration to show the politicians of the country that there was a genuine grass-roots, deeply-felt interest in the issue that crossed all political lines and all age groups. It was my conviction that nothing significant could be done until the politicians understood this. In other words, the issue had to become part of the political dialogue of the nation before we could hope to accomplish anything."[6]

Nelson was right—Earth Day spurred congressional action. On December 31, Congress passed the Clean Air Act of 1970—a sweeping rewrite of a 1963 law. The law established National Ambient Air Quality Standards (NAAQSs) for airborne pollutants, including ozone, lead, and carbon monoxide. It set technology standards for power plants and major industrial sources and strengthened limits on emissions from cars and trucks. Congress gave enforcement authority to the US Environmental Protection Agency, which had been formed earlier in the month. The 1970 law, with its amendments in 1977 and 1990, became the most extensive, and arguably the most important, environmental law in the country.

Also impressive was the staying power of Earth Day as an event. The first Earth Day was covered by major media outlets and was front-page news. Though subsequent Earth Days have not reached this media pinnacle or level of participation, the celebration of Earth Day has persisted over four decades, representing a political and public desire to protect the earth and its resources.

Nelson's Legacy

During his legislative career, Nelson sponsored or cosponsored a host of environmental bills, including the Wilderness Act of 1964 and the Wild and Scenic Rivers Act of 1968, which represented early legislative success. After the magic of that first Earth Day, Nelson saw major pollution control laws passed, which included the Clean Air Act, followed two years later by the Clean Water Act. He sponsored the Surface Mining Control and Reclamation Act of 1977, discussed in Chapter 4, which was meant to control the environmental damage caused by coal mining. He introduced the first bill to ban the pesticide DDT after the book *Silent Spring* by Rachel Carson was published in 1962.[17] (DDT was banned by the EPA in 1972—a topic that concerns the next hero, William Ruckelshaus.)

After serving eighteen years in the US Senate, Gaylord Nelson lost his reelection bid to Robert Kasten (R-WI) in 1980. He continued fighting for the environment as a counselor and member of the board for the Wilderness Society, a nonprofit environmental group. For these efforts alone, one could regard Nelson as an environmental leader. But it was envisioning and then inaugurating the first Earth Day that was perhaps his most heroic act. In 1995, President Bill Clinton honored Nelson with the Presidential Medal of Freedom, the nation's highest civilian honor. In giving the award to Nelson, President Clinton observed, "As the father of Earth Day, he is the grandfather of all that grew out of that event. He inspired us to remember that the stewardship of our natural resources is the stewardship of the American Dream."[18]

Nelson embodied the characteristics of an environmental hero: courage, determination, and wisdom. He courageously pursued environmental goals, especially when challenging Congress to pass tough environmental laws, and he was determined to make the first Earth Day a nationwide movement. And perhaps his greatest legacy was the wisdom he showed in capturing the energy of youth to make environmental issues part of a national conversation: the passion for environmental protection that he instilled in young people would influence public policies for decades.

Compelling, too, was his call to all Americans to live up to an environmental ethic. In this, he exemplified the need to make choices that advance the greater environmental and social good. In a speech presenting the environmental agenda of the first Earth Day to the Ninety-First Congress on January 19, 1970, Nelson noted that restoring the environment would require reshaping American values. "American acceptance of the ecological ethic will involve nothing less than achieving a transition from the consumer society to a society of 'new citizenship'—a society that concerns itself as much with the well-being of present and future generations as it does with bigness and abundance."[19] The yardstick by which American progress should be measured was a simple one, he proposed: a new assertion of "environmental rights and the evolution of an ecological ethic of understanding and respect for the bonds that unite the species man with the natural systems of the planet."[20]

Gaylord Nelson died in 2005 at the age of eighty-nine, still serving the Wilderness Society.

WILLIAM RUCKELSHAUS: FIRST AND FIFTH EPA ADMINISTRATOR

Throughout this book, we have seen evidence of what happens when government and regulatory agencies are lax in their duties. A strong EPA, with a leader who is committed to citizens and the environment over corporations, is key to ensuring that disasters like the ones covered in Chapters 2 to 5 are avoided or minimized. Our second hero is William Ruckelshaus, the first administrator of the EPA, and also the fifth. To understand William Ruckelshaus and his impact as an environmental hero, we must first consider the history of the EPA and the changing sentiment about a strong national presence in controlling industrial pollution.

The Political Backdrop of the Forming of the EPA

Though Presidents Kennedy and Johnson added environmental protection to their legislative agendas, Congress had resisted passing strong pollution control legislation in the 1960s. However, public support for environmental protection in the late 1960s and 1970s was high, as evidenced by the first Earth Day on April 22, 1970. Earth Day became the rallying call for Congress to act. Events such as the 1969 Santa Barbara oil spill that stirred Gaylord Nelson, an oil-slick fire on the Cuyahoga River (also in 1969), and the 1962 publication of Carson's classic *Silent Spring* on the devastating effects of pesticides had captured the attention of Americans. Congressional advocates, including Senators Nelson and Muskie, were pressing for a host of national environmental laws that would eventually be passed in the 1970s—often referred to as the first environmental decade.

Recognizing that the tide of public opinion favored environmental protection, President Nixon further extended the White House's environmental efforts. In signing the National Environmental Policy Act (NEPA) into law on New Year's Day 1970, he observed: "The 1970s absolutely must be the years when America pays its debt to the past by reclaiming the purity of its air, its waters, and our living environment. It is literally now or never."[21] A few weeks later, President Nixon would proclaim in his State of the Union message that the decade of the 1970s would be a period of environmental transformation. In December 1970, the EPA was created as an independent regulatory agency just weeks before the passage of the Clean Air Act—the first of a series of laws establishing national pollution control standards.

Nixon was less an environmentalist than a pragmatic politician. The EPA figured into his desire to reduce what he saw as a bloated federal bureaucracy by reorganizing federal agencies and combining functions. At the time, no single federal agency was devoted to environmental protection. Within the Executive Office of the President was the Council on Environmental Quality (CEQ), created as part of the just-passed NEPA. However, the CEQ's role was narrowly defined: it was there to serve as an environmental adviser to the president and to implement NEPA. Further limiting the reach of the CEQ was the fact that NEPA extended only to the actions of federal agencies. The chair of Nixon's Advisory Council on Executive Organization, Roy Ash, advocated for one regulatory agency solely directed to pollution control, rather than having that function dispersed among federal natural resource and public health agencies. Nixon agreed, and discussions about the role of the EPA began.

On July 9, 1970, as part of his reorganization plan, Nixon informed Congress of his wish to establish an independent agency with the goal of setting and enforcing pollution control standards. This fledgling agency would also conduct research on the effects of pollution, oversee and work with states through grants and technical assistance, and provide direction for new policies. Congress approved the EPA with little debate, as it was busy crafting the Clean Air Act of 1970, a major piece of legislation that established national air quality standards and would require a strong regulatory oversight agency.

Ruckelshaus Becomes the First EPA Administrator

On December 2, 1970, the EPA became a reality. Two days later, William Ruckelshaus took his oath as its first administrator, and Nixon signed the Clean Air Act on December 31. Things in the environmental policy arena were happening quickly, and it would take a remarkable leader to see these changes through successfully. That leader would be Ruckelshaus.

Though just thirty-eight years old at the time of his appointment to head the EPA, Ruckelshaus was widely regarded as a rising political star and someone who knew how to work with diverse interests—a skill he would need to guide this fledgling agency. Ruckelshaus had served as deputy state attorney general for the state of Indiana shortly after his graduation from Harvard Law School. While in that position, he drafted the Indiana Pollution Control Act, which was his only foray into environmental policy prior to leading the EPA. He won a Republican seat in the Indiana House of Representatives and served as majority leader in his first term. He then narrowly lost his bid for a seat in the US Senate, in 1968, to Birch Bayh (D-IN). Nonetheless, Ruckelshaus would soon be in public service at the national level. Nixon appointed him to serve as assistant attorney general for the Civil

Division of the US Department of Justice. It was Attorney General John Mitchell who would later recommend Ruckelshaus to Nixon as a nominee for the role of first EPA administrator.

The early years of the EPA were anything but easy, beginning with the gargantuan task of organizing the agency. Disparate activities from all over the federal bureaucracy had to be brought under the umbrella of the new agency. Ruckelshaus's charge was to bring together staff from fifteen different federal entities to form the nation's first environmental pollution control organization. That alone would have been a herculean undertaking, but Ruckelshaus had to accomplish this not only in the EPA headquarters in Washington, DC, but in ten regional EPA offices across the country, and to do so in a short amount of time.

Ruckelshaus would later recall:

> I needed to gain enough understanding of the nature of the agency, and what should be done, *before* organizing it, so that the organizational structure itself didn't get in the way of progress. By the same token, we needed to provide some structure in a timely fashion so that people didn't get discouraged and start drifting away from our central purpose. So in about four or five months—inundated with organization charts floating around my office—I just *chose* an organizational structure. It's been reorganized several times since, so obviously it wasn't a perfect structure. But it was important to provide some clear organizational framework.[22]

A second challenge involved making the agency work. The EPA's job, simply put, was to protect human health and the environment from risks due to pollution. But there was nothing simple about achieving that mission. Companies that had enjoyed essentially free rein in sending pollutants into the air, water, and land were now subject to national standards that had to be met. Complying with emission standards was something Corporate America was not accustomed to doing. State efforts to address pollution had typically been minimal prior to the first environmental decade because many state politicians were not eager to offend companies, especially those that were a major source of employment. States, it was argued, were engaged in what was called a "race to the bottom," where the more accommodating states were to polluting industries, the more likely it was that industries would stay or relocate there.[23] Companies that had long enjoyed these pollution "havens" were not eager to change their business practices for some new federal agency.

The EPA had to be tough enough to show that it intended to enforce regulatory requirements, regardless of the political backlash it might face. As Ruckelshaus observed, "It seemed to me important to demonstrate to the public that the government was capable of being responsive to their expressed concerns; namely, that we would do something about the environment. Therefore, it was important for us to advocate strong environmental compliance, back it up, and do it; to actually show we were willing to take on the large institutions in the society which hadn't been paying much attention to the environment."[24]

Ruckelshaus tackled industrial titans, including US Steel and Dow Chemical. He often met resistance from executives in major companies. When Ruckelshaus paid a courtesy visit to Ed Cott, CEO of US Steel, saying that environmental laws were serious and that the company's compliance was mandatory, Cott replied, "You know, we don't like you very much, and we certainly don't like your agency."[25] Ruckelshaus recalled that corporate executives viewed pollution control laws as "an interference with an important part of the whole economy of the country; that once people understood this, this current fad having to do with the environment would go away and they'd be left alone again to do what they wanted."[26] Ruckelshaus was determined not to let the "fad" of support for environmental efforts fade, even if it meant that he would be rebuked by politicians who felt that the regulations were unwarranted or too strict.

One highly controversial early action taken by Ruckelshaus was to ban the widely used pesticide DDT (dichloro-diphenyl-trichloroethane). Responsibility for regulating pesticides had been transferred from the US Department of Agriculture (USDA) to the EPA as part of Nixon's reorganization plan. The USDA had prohibited some uses of DDT, but had not canceled the registration of the pesticide. When the EPA assumed regulatory authority for pesticides, it issued notices of intent to cancel all remaining federal registrations of products containing DDT. Over thirty DDT formulators requested a hearing to avoid cancellation. They found a receptive audience in Edmund M. Sweeney, the hearing examiner, who oversaw seven months of hearings regarding the health and environmental effects of DDT. In his ruling, Sweeney observed that "DDT is not a carcinogenic, mutagenic, or teratogenic hazard to man. The uses under regulations involved here do not have a deleterious effect on fresh water fish, estuarine organisms, wild birds, or other wildlife."[27] Sweeney found that the pesticide provided essential benefits to agriculture, and that the current restrictions on DDT use were appropriate.

Environmentalists were outraged, while agricultural interests were pleased. But Sweeney's decision would not be the last word. Ruckelshaus overturned Sweeney's findings and banned DDT, effective December 31, 1972, saying that the "long-range risks of continued use of DDT . . . were unacceptable and outweighed any benefits."[28] The pesticides industry filed suit to nullify the EPA ruling, but was unsuccessful.[29] The decision remained controversial for years as critics argued that Ruckelshaus had overstepped his authority. Ruckelshaus, however, saw the DDT ban as a demonstration of the willingness of the EPA to step up to its responsibility to do the right thing.

A third challenge was establishing effective intergovernmental relations. State and local governments did not appreciate the intervention of this new agency that mandated changes in the practices of businesses within their borders. Moreover, national environmental laws placed new requirements on state and local facilities such as municipal wastewater treatment plants. Ruckelshaus's enforcement of regulatory requirements on state and local governments did little to endear him to his fellow public servants. To prove that he was serious about enforcement, Ruckelshaus and EPA staff announced at a national mayors' conference that the agency would sue Atlanta, Cleveland, and Detroit for sewage treatment violations under the Clean Water Act. This public shaming of local governments that had ignored treatment requirements brought cries from state and local officials that the EPA was overstepping its authority. Nonetheless, the law was enforced.

Enforcement of these new environmental requirements was not all that bothered state and local officials; state officials were also irked by intergovernmental legal relationships. Laws such as the Clean Air Act and the Clean Water Act were based on a "partial preemption" approach, meaning that state programs were usurped by federal standards. States could reassume regulatory control under air, water, and hazardous waste programs only after adopting standards at least as stringent as the national ones, and then only with EPA approval. Taking away state authority over environmental programs until the state adopted regulations similar to or stronger than those set by the EPA did not sit well with many state environmental officials. They strongly resented the EPA's dictating what standards would be enforced by state environmental agencies. By many accounts, state-EPA relationships were terrible.[30] Ruckelshaus made a point to visit state environmental officials around the country to try to explain the nature of this new federal oversight, but he often was met with resentment from state officials who were reluctant to take a backseat to the EPA. Nonetheless, he persisted in trying to gain the trust of state officials.

A fourth challenge was working with the presidential administration. Even though Nixon had worked to establish the EPA, the Oval Office resented the agency's encroaching on corporate and government actions. Nixon had campaigned on a "new federalism" that would give states more autonomy, and EPA oversight of state environmental efforts certainly did not fit that model. Corporate America had the president's ear, and Nixon advised Ruckelshaus to not be pushed around by the "crazies" in the new agency who were eager to compel compliance and write tough regulations. Moreover, Nixon was a reluctant environmentalist: he favored environmental protection efforts only because to do so was politically popular. Nixon created the agency, as Ruckelshaus remembers, not because he had concerns about environmental protection, but because people were demanding that the national government act.[31]

Integrity in Public Service: Watergate

In three short years at the helm of the new agency, Ruckelshaus had demonstrated courage, levelheadedness, and integrity. He had gone toe-to-toe with industries challenging EPA authority, with state and local governments that were unhappy with EPA oversight, and with a presidential administration eager to appease American companies. Ruckelshaus would subsequently cross swords with Nixon in a far more public way during the Watergate scandal.

Nixon handily won reelection in November 1972, taking more than 60 percent of the popular vote. However, what had transpired during the campaign would traumatize the country. Members of Nixon's reelection team conducted a massive spying effort against his Democratic opponent, Senator George McGovern, and the Democratic National Committee (DNC). In the summer of 1972, five men who would eventually be linked to the Nixon campaign team were caught trying to bug the offices of the DNC at the Watergate Hotel and office complex.

A congressional investigation of the scope of the Watergate crimes and subsequent cover-up was under way in 1973, with hearings and news coverage of possible criminal activity within Nixon's administration. At Nixon's request, Ruckelshaus resigned from the EPA in April 1973 to serve as acting director of the FBI after Patrick Gray, then acting director of the FBI, testified to his role in destroying documents related to the Watergate investigation. During his two-month service as acting director, Ruckelshaus found wiretap records that had been "lost" at the FBI, and he also became convinced that the Watergate scandal went all the way to the White House. That conviction put him in the crosshairs of the president's inner circle of advisers.

Later that year, Ruckelshaus become US deputy attorney general, a position that put him directly in conflict with Nixon's attempt to contain the ever-growing Watergate scandal. Ruckelshaus would serve only twenty-three days. He resigned on October 20, 1973, together with US Attorney General Elliot Richardson, rather than carry out Nixon's order to fire special prosecutor Archibald Cox in what became known as the Saturday Night Massacre. Nixon fired him and Richardson that night, prompting Ruckelshaus to later quip, "Depending on your point of view, I was both fired and resigned."[32] Ruckelshaus would later observe, "When you accept a presidential appointment, you must remind yourself that there are lines over which you will not step—lines impossible to define in advance but nevertheless present. The line for me was considerably behind where I would have been standing had I fired Cox. In this case, the line was bright and the decision was simple."[33] Facing impeachment, Nixon resigned in August 1974. *Time* magazine would later list Ruckelshaus as one of the "Top 10 Best Cabinet Members," noting that he not only stood up to Nixon during the infamous Saturday Night Massacre but was also "no pushover at the EPA either."[34]

Integrity in Public Service: Repairing the EPA

After the Saturday Night Massacre prompted his resignation from the Nixon administration, Ruckelshaus eventually returned to the private sector. He accepted a position as senior vice president for Weyerhaeuser, a major lumber and timber company. Bruised by the Watergate affair, Ruckelshaus might have preferred to stay in this position for many years, but, once again, he was asked to deal with a scandal—this one rocking the integrity of the very agency he had worked to establish. The EPA was under fire, not for unflinchingly enforcing environmental laws, but for kowtowing to industrial interests.

Anne Gorsuch Burford, who had been selected in 1981 for the EPA's top post by President Ronald Reagan, soon came under fire for her allegiance to polluters and reluctance to implement the hazardous waste and cleanup laws. Burford, a conservative lawyer and Republican former state legislator from Colorado, was ideologically aligned with Reagan's pro-business, antiregulatory agenda. The Reagan administration sought to reverse the growth of the national government and federal regulations, including those of the EPA. More than half of the regulations targeted for reform by the

Reagan administration were EPA rules, a target supported by the new EPA administrator.[35] Burford also believed that the agency was overstaffed, and she submitted a budget proposal that cut over four thousand employees.[36] Under her leadership, EPA enforcement actions dropped by 60 percent, and the budget was cut by nearly 30 percent.[37]

Even more troubling was the inaction surrounding the implementation of new hazardous waste and cleanup regulations. In 1980, Congress had passed the Comprehensive Environmental Response, Compensation, and Liability Act (CERCLA). As described in Chapter 5, the so-called Superfund law required identification and cleanup of America's most toxic sites. Burford, EPA assistant administrator Rita Lavelle, and other Reagan administration EPA appointees were making decisions about the implementation of the new law, including how funds dedicated to cleanup should be spent. Lavelle oversaw the agency's newly established cleanup trust fund, the $1.6 billion Superfund.

As charges of impropriety and political favoritism grew, Congress launched an investigation into the actions of the EPA, demanding documents that it believed would shed light on how the agency operated. Burford, following Reagan's claim of executive privilege, refused to cooperate with Congress, making her the first agency director in US history to be found in contempt of Congress. Burford resigned on March 9, 1983, leaving in her wake a disgraced agency and demoralized employees in what is regarded as the largest scandal in EPA history. Though she insisted that she was a scapegoat for the wrongdoings of her boss and the White House, Lavelle was convicted of perjury and obstructing a congressional investigation, fined $10,000, and sentenced to six months in jail.[38] Over twenty Reagan administration EPA appointees resigned under pressure, leaving the agency in a kind of leadership limbo.

Ruckelshaus's service as the first EPA administrator, as well as his courage under fire during the Watergate scandal, made him a candidate to restore the reputation and morale of the EPA as its fifth administrator. Though some environmental groups worried that Ruckelshaus's position at the Weyerhaeuser Company would influence his willingness to enforce environmental laws, his nomination was confirmed by the US Senate less than three weeks after Burford's resignation in May 1983.

During his second term as EPA administrator, Ruckelshaus worked to restore the agency's budget and revitalize the enforcement program, but perhaps most importantly, he gave employees a signal that the agency would act with integrity and environmental purpose. At the time of his appointment, over seventeen thousand abandoned or inactive hazardous waste sites had been identified. Ruckelshaus worked to increase the Superfund budget by more than 50 percent and ramped up efforts to recover cleanup costs from responsible parties.

Ruckelshaus would remember this as a time to put the EPA back on an even keel, to restore the trust of the public, and to begin to rebuild morale inside the agency.[39] A *New York Times* editorial would put it more bluntly: "William Ruckelshaus healed the Environmental Protection Agency that Anne Burford shattered because he was trusted on all sides. In his too-brief tenure he restored the professionalism and quality he had given the agency as its first Administrator."[40]

Ruckelshaus's Legacy

Ruckelshaus served as the EPA's fifth administrator until January 1985. After leaving the agency, he served on the World Commission on Environment and Development (the Brundtland Commission), where he had an opportunity to influence the direction of sustainable development at an international level. Ruckelshaus also served on a number of corporate boards and participated in several university initiatives, including the Washington State University Policy Consensus Center. In 2001, President George W. Bush appointed him to the US Commission on Ocean Policy.

Ruckelshaus had a reputation as a man of integrity who was also a consensus-builder. He believed in the value of collaboration and in the power of democratic values to inform environmental protection efforts. In his 2005 keynote speech to the National Council for Science and the Environment, Ruckelshaus urged: "We need to identify problems, be guided by the best science, and not fear to act, because through action we learn. If our action is ill advised or fails, then adapt, adjust our system, and try again. . . . One thing we must do if our form of democracy is to work is to holster our political guns and lower our voices. After all, it's easier to listen with our mouth closed."[41] He emphasized that collaborative processes must, "in the words of Donald Snow of Montana's Northern Lights Foundation, 'break through the shallow facade of rhetoric and reach to the heart of the issue.' Only then, when people are united despite their differences by hard-earned trust, does the astounding political power of collaboration become effective."[42]

The William D. Ruckelshaus Center at Washington State University is a testament to collaborative environmental solutions in the Pacific Northwest. In 2015, Ruckelshaus was awarded the Presidential Medal of Freedom (about a decade after Nelson received the same honor). In announcing the award, the White House noted that Ruckelshaus was a dedicated public servant who had shaped the guiding principles of the EPA and worked diligently to bring the public into the decision-making process.

In sum, Ruckelshaus was an environmental hero. His choices were ethical, his compass was directed to the greater good, his work at the EPA was widely recognized as fair, and he acted with integrity. As of this writing, he continues to build on his belief that people working together, collaboratively, can effect positive change for a brighter future.

JUDY BONDS: APPALACHIAN ACTIVIST

Gaylord Nelson founded Earth Day, and William Ruckelshaus set the direction for the EPA. Julia "Judy" Bonds, our third hero, held no high office, nor did she serve in the government. A coal miner's daughter, Judy Bonds was born and raised in the Appalachian Mountains of West Virginia. As part of the seventh generation in her family to grow up in Marfork Hollow (also referred to as "holler"), Bonds had no plans to leave and envisioned staying in her ancestral home. However, Marfork soon felt the ravages brought about by mountaintop removal coal mining.

As described in Chapter 4, mountaintop removal is the practice of blasting off the top of a mountain to reach the coal beneath. When done irresponsibly, valleys are filled with mining overburden, streams are polluted, forests are destroyed, and gigantic slurry dams thick with heavy metals may cause flooding, contaminating drinking water and threatening homes. Homes near these enormous operations may be damaged by the blasting that continues unabated while coal seams are exposed. Residents face an increased incidence of asthma and other respiratory diseases owing to their exposure to coal dust and airborne toxins. Other adverse health effects linked to mountaintop removal coal mining include traumatic stress and increased incidence of lung cancer.[43]

Chapter 4 laid out the ill effects of coal mining, which Judy Bonds would come to know in a very personal way. Not only did coal mining force her from her home, but she believed that it cost her her life.

From Waitress to Environmental Hero

Bonds was from a coal mining family: her father, grandfather, and ex-husband, as well as other family members, had mined coal from nearby coal seams not far from Marfork. She understood that coal mining was a necessary though dangerous and sometimes deadly way of life in Appalachia; her father had worked the mines until shortly before he died of black lung disease. Still, she was unprepared for what Massey Energy "brought down on our heads in Marfork."[44] The practice of mountaintop removal coal mining created more waste than coal and covered the town in coal dust. Her grandson developed asthma, and her home and neighbors' homes were made unbearable by the showers of coal dust and debris. As they watched neighbors leave the holler, Bonds and her family retained a lawyer to fight Massey. She lost the battle to keep King Coal from decimating the holler and eventually sold her home to Massey in 2001. The last family to leave the community of Marfork, the Bondses moved nine miles away to Rock Creek, West Virginia. The town of Marfork, like other Appalachian towns, was then deserted. The nearby mountaintop removal operation had turned it into a wasteland, and now it served as a constant reminder of the high cost of mountaintop coal mining paid by the residents of Appalachia.[45]

Bonds was proud of what she often called her "hillbilly heritage" and loved the mountains of West Virginia. Even as she saw the land surrounding her home destroyed, Bonds might well have just accepted the seemingly unstoppable rush to mountaintop coal mining and kept her job as a waitress and manager at the local Pizza Hut. But as Bonds would describe it, she had an environmental epiphany. She watched her grandson as he stood in a blackened, polluted stream in the Coal River Valley, with his fists full of dead fish. He looked at his grandma and asked what was wrong with the fish. Bonds would later recount: "I looked down at the water and screamed. My family, for generations, has enjoyed that stream, but we never went back in the river again."[46]

Her grandson also told her of an escape route from his school he had planned should a massive coal waste dam break and flood their valley. The 924-foot-high Brushy Fork Impoundment was an earthen dam owned by Marfork Coal, a subsidiary of Massey Energy, that sat three miles above Bonds's ancestral home and was permitted to eventually hold nine billion gallons of waste.[47] If it were to collapse, a 40-foot wall of sludge would engulf communities as far as fourteen miles away.[48] Bonds would recall that moment: "I knew in my heart there was really no escape [from the dam]. How do you tell a child that his life is a sacrifice for corporate greed?"[49] The potential peril that coal mining posed for her grandson and the sight of the fish killed in the stream were the catalysts that turned Judy Bonds into an environmental hero.[50]

Though Bonds never called herself a hero, she did have an arch-nemesis: Massey Energy. For Bonds, this was personal. Massey Energy owned Marfork Coal, the coal company that moved into Marfork Hollow. As the largest coal producer in central Appalachia at the time, Massey owned most of the coal deposits and dozens of smaller mining operations.[51] Massey and its controversial CEO, Don Blankenship, saw mountaintop removal mining as an expedient and cost-effective way to get at coal seams located deep underground. Blankenship was used to getting what he wanted. With political connections, economic clout, and the ear of state and national politicians, Massey was the epitome of King Coal in West Virginia. But King Coal was about to meet a fierce adversary.

Two weeks after Bonds and her grandson witnessed the fish kill, she saw a flyer on a window promoting a rally against irresponsible mining. She went to a meeting and then a rally, and in her words, she "never looked back."[52] Bonds joined the few local residents brave enough to stand up to the coal industry and its practice of mountaintop removal. She would spend the rest of her life—more than a decade—fighting the coal industry, and Massey in particular. It proved to be a tough and dangerous battle. Even though mining practices destroyed homes, increased flooding, and sheared the tops off of their beloved Appalachian Mountains, most residents remained beholden to what was their community's major source of income. Many West Virginian counties relied on coal mining, and residents were willing to ignore the human health risks and environmental damage in local areas so deeply reliant on King Coal. The more vocal she became, the more unwelcome Bonds was in her Appalachian homeland.

Coal River Mountain Watch, a fledgling grassroots environmental group, offered Bonds a place to champion her fight and a chorus of voices to stand up to reckless mining practices. Bonds volunteered for the group in 1998, shortly after her experience with her grandson and the fish kill. The group, dedicated to fighting mountaintop removal coal mining, had started that year with only a few volunteers who had pledged to fight for "social, economic, and environmental justice."[53] Three years later, Bonds became the organization's paid ($12,000 a year) outreach director.[54] This position offered Bonds a platform from which to organize protest rallies, engage in letter-writing campaigns, file lawsuits, and testify at hearings. The newly formed grassroots organization operated, however, on a shoestring budget—hardly the funding needed to confront the multibillion-dollar coal industry.

Bonds made many personal sacrifices in going against Massey and other coal mining companies. She faced many threats and was insulted, intimidated, and arrested during protest rallies. During a peaceful march against mountaintop removal, a coal miner's wife rushed through the crowd, found Bonds, and slapped her hard across the face. Captured on a cell phone and uploaded to YouTube, this incident illustrated the kind of daily harassment faced by anyone who dared to take on King Coal. Bonds also received threatening calls in the middle of the night, as did others in the organization.

And that was not the worst of it. In an interview, Bonds lamented, "You really haven't been intimidated until you see a 60-ton coal truck swerve at you on a narrow road, when there's a rock cliff on one side and a 100-foot drop-off on the other. I have a friend that says the only difference between now and the 1920s [when coal companies persecuted and even killed union organizers] is that they're not shooting us on courthouse steps. They're running us over with coal trucks."[55] This was not hyperbole, as coal trucks routinely carried a heavier load than was legally allowed, an especially dangerous practice on narrow and steep highways. Over a two-year period, fourteen people were killed in accidents with overweight coal trucks.[56] Bonds forged a partnership with the United Mine Workers union to change this illegal practice. Working with the union and other activists, Bonds filed a lawsuit against coal operators to force them to carry safe and legal loads. She launched a national grassroots campaign asking people to send postcards to West Virginia's governor with a pledge that they would not visit the state until the oversized trucks were off the highway.[57]

Bonds also spoke out against violations of federal and state mining laws. Its attention called to one such violation, the West Virginia Mining Board suspended a Massey-operated coal mine for thirty days, prompting the ire of mine managers. Bonds was threatened by armed guards at Massey when she sought to show people the devastation of mountaintop removal. Some of these individuals were journalists, who wrote up stories about the destruction in Appalachia in national papers. As media attention was kindled, Massey launched a massive ad campaign to improve its image and tried to keep activists from picketing its annual meeting.[58]

Bonds testified at many local, state, and national level hearings, where she was vilified by coal men and women who saw her as a threat to their way of life. Coal companies, as Bonds told it, would pack local permit hearings with their workers, who were told that environmental activists would take their jobs away. Once she heard someone at a permit hearing warn, "If I were these ladies [environmental activists], I'd be afraid to go home tonight." To which Bonds replied, "Well, you can't scare Appalachian women, and they ought to know that. We've had to fight all our lives."[59]

From the Holler to National Renown

Her courage soon found a national stage. On a January evening in 2003, Bonds received a phone call. Richard Goldman was calling to let her know that she was the North American recipient of the Goldman Environmental

Prize. She was awarded one of six Goldman Environmental Prizes on April 14, 2003, for her work with the Coal River Mountain Watch. The $125,000 prize, sometimes referred to as the Nobel Prize for activism, annually recognizes six environmental heroes throughout the world. With her award money, she paid off her mortgage, helped her daughter buy a car and her grandson's braces, and then donated nearly $50,000 to Coal River Mountain Watch, an amount equal to the organization's annual budget.[60] When interviewed about receiving the prestigious award, Bonds commented that now "we can expose this secret of what's been going on in the coalfields of Appalachia."[61]

True to her word, Bonds used the attention created by her award to focus media attention on the plight of the communities surrounded by mountaintop removal coal mining operations. The spotlight on Appalachia in the wake of Bonds's receipt of the Goldman award prompted reporters to go to West Virginia and witness firsthand the dangers and destructive power of this type of mining.

Bonds and fellow activists would need this kind of media attention, and more. Coal River Mountain Watch, together with other grassroots environmental organizations, would soon enter a national political arena when they challenged the actions of the Bush administration. Shortly after taking office in 2001, President George W. Bush, a former oilman, and Vice President Dick Cheney, past CEO of Halliburton, wasted little time establishing a national energy policy that promoted domestic drilling of oil and the mining of coal. The Bush administration saw the need to support the country's fossil fuel industry as a national security issue following the 9/11 terrorist attack, and its energy policy task force advocated a national energy plan heavily reliant on continued coal production to preserve coal's preeminence as the country's biggest source of electricity (see Chapter 4). As President Bush observed at a meeting of the West Virginia Coal Association, "Coal is affordable and coal is available right here in the United States."[62]

Perhaps most egregiously, the federal government allowed coal mines to engage in the practice of "valley fill"—dumping mining overburden (rock and dirt from the decapitated mountain) into valleys. The US Army Corps of Engineers issued these valley fill permits under authority given to it in Section 404 of the Clean Water Act. Though the Corps had granted permits in previous administrations, controversy surrounding such large disturbances of land soon galvanized environmental groups. Activists sued the Corps, winning in federal district court. US District Judge Charles Haden found that valley fill permits, which allowed the dumping of huge amounts of mining overburden into valleys and streams below the mountaintop mining operation, were illegal under the Clean Water Act. Haden's ruling stated that the dumping of mining overburden was the same as dumping waste, and not at all the same thing as the dumping of dredged and fill materials from waterways typically permitted by the Corps. Though Haden's decision was eventually overturned on appeal, the Bush administration nonetheless decided to broaden the interpretation of the dredge and fill material rule. The new rule would allow materials like mining overburden to be placed in valleys and streams as long as doing so was "associated with an appropriate project."[63] Appropriate projects included mountaintop mining.

Daunted but still determined, Bonds and other activists continued to fight the battle against mountaintop mining. When Bonds became codirector of Coal River Mountain Watch in 2007, she gained even more opportunities to testify in hearings and to represent environmental interests in West Virginia. That year she testified before the House Natural Resources Committee, arguing that the US Office of Surface Mining Reclamation and Enforcement (OSMRE) consistently allowed state environmental agencies to grant variances exempting coal companies from requirements to restore the land to its approximate original contour, a requirement under the Surface Mining Control and Reclamation Act of 1977 (the bill that Gaylord Nelson had sponsored). Under the law, coal companies are required to restore mined lands to their original configuration through backfilling and grading. Massey and its subsidiary coal companies made no attempt to restore mountaintops to their former contour.

In 2008, Bonds was among those who fought another pro–mountaintop mining rule change. This time it was the Bush administration's relaxation of the stream buffer zone rule in the waning days of its second term. This rule, originally promulgated by OSMRE in 1983, prohibited mining operations within one hundred feet of a stream. Coal operators could obtain waivers, but they had to show that mining operations would not adversely affect water quality. Though state and federal regulators had granted exceptions to coal companies in the past, the rule change under the Bush administration gave coal operators virtually free rein to bury streams with mining overburden. Led by Coal River Mountain Watch, Appalachian environmental groups fought the rule change, which would go into effect on January 12, 2009—just days before a new administration took office. They sued the Department of the Interior for violations of the Clean Water Act. They argued that relaxing the rule further would only hasten the destruction of Appalachian streams. At the time, over two thousand miles of streams had been buried by mining waste.[64]

Newly elected President Obama and members of his administration were amenable to making changes favored by environmentalists to regulate mountaintop removal coal mining. In April 2009, Ken Salazar, the new secretary of the interior, called the stream buffer rule created by the Bush administration a "major misstep" and "bad public policy." He asked that the rule be vacated by the courts and sent back to the Department of the Interior for further action. The proposed rulemaking temporarily halted the lawsuit. But Bonds wanted to ensure that real changes would be made, noting, "We're happy the administration is realizing the error of the Bush rule, but just being better than Bush is not OK. . . . We want to know if the [1983] stream buffer zone rule will really been enforced."[65]

On July 24, 2009, after the Obama administration promised to reform mountaintop removal mining but not to eliminate it, environmentalists swung into action. Over two hundred activists gathered at Marsh Fork Elementary School, located adjacent to a 2.8 billion–gallon toxic coal sludge impoundment, to advocate the relocation of the school and to protest irresponsible mining. Activists marched toward the Massey Energy property, facing hundreds of angry coal-mining supporters who taunted the activists with air horns and yelled, "This is our state."

Police arrested Bonds, along with more than two dozen nonviolent protesters, for trespassing on Massey property. Among those arrested were National Aeronautics and Space Administration (NASA) scientist James Hansen, actress Daryl Hannah, and Ken Hechler, a former Democratic US representative for West Virginia; these high-profile arrests garnered more national attention to the environmentally destructive mining.[66] Not only had miles of mountain waterways been damaged or destroyed and 400 mountaintops flattened, but 1.4 million acres of forest would be at risk if mountaintop removal mining was allowed to continue.[67]

Later that year, when the Obama administration issued a Memorandum of Understanding among the Corps, the Department of the Interior, and the EPA to toughen federal rules governing mountaintop removal, Bonds was still not satisfied. She commented, "What I'm seeing so far is basically no change whatsoever—yet. It just looks a lot like smoke and mirrors to me. It seems like this administration is saying that we're going to look harder at these permits before we rubber-stamp these permits."[68] Her skepticism was warranted. The relaxed stream buffer rule, a product of the Bush administration, remained in place until 2014, when a federal court found that it violated the Endangered Species Act.[69] It would take until December 19, 2016, in the waning days of the Obama administration, until OSMRE published strict new guidance restricting the dumping of coal waste into streams in its stream protection rule.

In the up-and-down nature of politics, a Republican-controlled Congress passed a law just two months later that repealed the newly issued stream protection rule, which President Trump signed. Trump was pleased to sign the bill, noting that "another terrible job-killing rule" had been eliminated and that ending it would save "many thousands of American jobs, especially in the [coal] mines, which [is something] I have been promising you."[70] This was one of several efforts by the newly elected Trump administration to support the coal industry.

After nearly a decade fighting for responsible coal mining practices, Bonds was well known among environmentalists, local residents, and local and state mining officials. She had become skilled at interacting with members of Congress, lobbying for state and national laws, and working with federal and state regulators, but perhaps most important, she had become an inspirational speaker. Environmental organizations around the country invited her to speak at their events. She accepted many of these invitations and would be on the road and away from her beloved West Virginia mountains for months at a time. In her speeches, Bonds minced no words in describing the damage done by coal companies such as Massey. She compared southern West Virginia to a "war zone," noting that "three and one-half million pounds of explosives are being used every day to blow up the mountains. Blasting our communities, blasting our homes, poisoning us, trying to intimidate us. I don't mind being poor. I mind being blasted and poisoned."[71]

Nor did she hesitate to identify who needed to step up to the challenge of environmental protection and be the force for change. At the first meeting of Power Shift, a youth summit on climate change policy, Bonds challenged the six thousand youth in attendance to get involved. "Arm yourself with knowledge, arm yourself with truth. Everyone has a place in this movement. Find your place. Make being an environmental activist cool and sexy as though your life depended on it. Because your life *does* depend upon it. Remember that green is the only color that matters."[72]

In 2009, she joined thousands of others at a mass rally to influence Congress to act on climate change. A keynote speaker at the Capital Climate Action rally in Washington, DC, Bonds urged college students to act, saying, "Not only are you changing America, the youth standing here today are the ones who will change the world."[73] In June 2010, less than two months after the Upper Big Branch coal mine explosion killed twenty-nine miners, activists again battled Massey Energy over its mining practices, prompting the arrest of more than one hundred members of the activist group Climate Ground Zero.

Bonds's Legacy

Judy Bonds died on January 3, 2011, of cancer at the age of fifty-eight. Her death was reported in major news outlets, including the *New York Times,* the *Washington Post,* and the *Boston Globe.* Even more than Nelson and Ruckelshaus, Bonds had grit. "The thing about Judy, she never backed down from anything," said Vernon Haltom, codirector of Coal River Mountain Watch. "That's the kind of courage she had and the kind of courage that she needed to stand up to great odds with only her courage and conviction to back her up."[74] Ken Hechler may have best captured the courage of Judy Bonds. In 1999, he recalled, he organized a march with Bonds to commemorate the 1921 Battle of Blair Mountain, when coal miners attempted to unionize. Soon the group of marchers were surrounded by angry counterdemonstrators. "We were attacked by a group of toughs. And I looked over and I saw Judy Bonds, and she had great determination on her face. I started out being scared, and then terrified, but I got inspired by her courage."[75]

Equally compelling was her belief in people working together as a positive force for change. Calling Bonds the "patron saint of the anti–mountaintop removal movement," Jason Howard, author of *Something's Rising,* observed of her life: "The old saying goes that you don't know what you've got till it's gone. But luckily, I don't think that ever applied to Judy. The environmental movement knew her value while she was still with us. We knew that she was a remarkable force. We knew when she approached a speaker's podium what was about to happen. We knew that she would be speaking truth to power until the very end. And that's exactly what happened."[76]

One might think that Bonds did not have time to become an environmental hero. Her environmental career did not span decades, as Gaylord Nelson's and William Ruckelshaus's did. But she accomplished more during her time at Coal River Mountain Watch than many others who have decades to pursue a cause. Her homespun opposition to mountaintop removal brought national attention to this egregious practice and inspired countless others to get involved. Bonds had a favorite saying: "*You* are the ones you've been waiting for." What she meant was that it is up to individuals to fight for what they believe in.[77] Bonds lived that belief. Her dedication, straight talking, and seemingly boundless energy quickly made her one of the nation's leading community activists.

CONCLUSION

Gaylord Nelson, William Ruckelshaus, and Judy Bonds were very different individuals, with different backgrounds, education, and life experiences. Still, each was a hero. What are the common elements that can be gleaned from their stories?

First, all three exhibited courage in the face of sustained opposition. For Nelson, it was a recalcitrant Congress reluctant to enact tough pollution control laws. Ruckelshaus faced business interests that sought to weaken the newly created EPA as he struggled with a president who was ill inclined to move vigorously against large corporations. Bonds faced down the massive, well-entrenched coal industry in West Virginia, even at great personal cost.

Second, each had a persistent optimism. Nelson's hopes that President Kennedy's conservation tour would prompt an environmental revolution were dashed, as the presidential tour drew anemic crowds and little media interest. Ruckelshaus, after reluctantly agreeing to resign as EPA administrator to assist the Department of Justice during the Watergate scandal, faced another presidential firestorm when he was asked to return to the agency following charges of inappropriate conduct by EPA administrator Anne Gorsuch Burford. Bonds's story—perhaps the most poignant—was of a battle with major coal companies, an administration wedded to the coal industry, and her neighbors who sought to protect coal mining as their livelihood. Each of these three heroes faced seemingly insurmountable odds, but each faced those odds with a belief that they could accomplish what they needed to do.

Third, these environmental heroes armed themselves with knowledge and a deeper understanding of the need for environmental protection. Nelson visited the site of the Santa Barbara oil spill; Ruckelshaus learned about DDT and a host of chemical contaminants; Bonds came to understand the political process as well as the scientific data on the effects of coal mining. Each spoke from a position of deeper knowing.

Fourth, and most important to any environmental story, each hero embraced the importance of civic engagement. Without Earth Day and Nelson's faith that this event would catapult environmental protection to the top of the congressional agenda, American interest in protecting the environment might have languished for many more years. The first environmental decade might have come much later than the 1970s, or not at all, if not for the energy brought by millions of people to that first Earth Day on April 22, 1970. It seems fitting that Earth Day 2017 was marked by marches around the country protesting the new administration's devaluation of science and the rollback of various environmental rules—including the ones Bonds and her allies fought so hard to get. Ruckelshaus's successful tenure at the EPA depended in no small way on the continued support of citizens who demanded that the national government move forward to clean the air, the water, and the land. Bonds's belief in people was evident in all that she did—from building a grassroots environmental organization to rallying thousands of people to engage in nonviolent action.

The next chapter explores ways in which each of us can exhibit courage, persistence, optimism, and the belief that together we can accomplish great things.

DISCUSSION QUESTIONS

1. Which of the three individuals described in this chapter seems most heroic to you?
2. Explore the list of hero characteristics. Which characteristic is most compelling? Are any characteristics missing from this list?
3. What did Judy Bonds mean when she said, "You are the ones you've been looking for"?
4. The chapter emphasizes the role of individual heroes. Is it possible for organizations to act heroically? Why or why not?
5. What was it about William Ruckelshaus that made him a candidate to return as EPA administrator in the Reagan administration?
6. Why do you think the Earth Day event has persisted over the decades? Why have subsequent events failed to attract as many participants as the first Earth Day?

NOTES

1. Sanjoy Hazarika, "Indian Journalist Offered Warning," *New York Times,* December 11, 1984.

2. Raajkumar Keswani, "An Auschwitz in Bhopal," *Bhopal Post,* June 28, 2010, www.thebhopalpost.com/index.php/2010/06/an-auschwitz-in-bhopal/ (accessed December 11, 2010).

3. Keith Schneider, "Gaylord A. Nelson, Founder of Earth Day, Is Dead at 89," *New York Times,* July 4, 2005.

4. Gaylord Nelson, with Susan Campbell and Paul Wozniak, *Beyond Earth Day: Fulfilling the Promise* (Madison: University of Wisconsin Press, 2002), 164.

5. Gaylord Nelson, personal communication to President John F. Kennedy, August 29, 1963, in correspondence between Kennedy and Nelson, Gaylord Nelson and Earth Day, The Nelson Collection, http://nelsonearthday.net/docs/nelson_231 -16_nelson-kennedy_correspondance_re_trip.pdf.

6. Nicholas Lemann, "When the Earth Moved: What Happened to the Environmental Movement," *The New Yorker,* April 15, 2013.

7. Quoted in Schneider, "Gaylord A. Nelson, Founder of Earth Day, Is Dead at 89."

8. Nelson, Campbell, and Wozniak, *Beyond Earth Day,* 170.

9. Miles Corwin, "The Oil Spill Heard 'Round the Country," *Los Angeles Times,* January 28, 1989.

10. Gaylord Nelson, letter to CBS president Frank Stanton, April 7, 1971, Gaylord Nelson and Earth Day, The Nelson Collection, http://nelsonearthday.net/docs /nelson_2-15_CBS_news_letter.pdf.

11. Earth Day Network, "The History of Earth Day," www.earthday.org/about/the -history-of-earth-day/ (accessed September 24, 2017).

12. Ibid.

13. March for Science, https://satellites.marchforscience.com/ (accessed June 23, 2017).

14. "Earth Day—1970: Mass Movement Begins," *The Gaylord Nelson Newsletter* (May 1970), Gaylord Nelson and Earth Day, The Nelson Collection, http://nelsonearthday.net/images/nelson_newsletter_may70.jpg.

15. "Partial Text for Senator Gaylord Nelson, Denver, Colo., April 22," Gaylord Nelson and Earth Day, The Nelson Collection, http://nelsonearthday.net/docs /nelson_26-18_ED_denver_speech_notes.pdf.

16. Nelson, letter to Stanton, April 7, 1971.

17. Bill Christofferson, "Nelson, Gaylord," 2014, American National Biography Online, www.anb.org/articles/07/07–00844.html (accessed June 24, 2017).

18. Quoted in Schneider, "Gaylord A. Nelson, Founder of Earth Day, Is Dead at 89."

19. Nelson, Campbell, and Wozniak, *Beyond Earth Day,* 174.

20. Ibid.

21. "The Guardian: Origins of the EPA," EPA Historical Publication 1 (Washington, DC: EPA, Spring 1992, updated September 6, 2016), https://archive.epa.gov/epa /aboutepa/guardian-origins-epa.html.

22. Michael Gorn (interviewer), "William D. Ruckelshaus: Oral History Interview," EPA 202-K-92-0003 (Washington, DC: EPA, January 1993, updated September 7, 2016), https://archive.epa.gov/epa/aboutepa/william-d-ruckelshaus-oral -history-interview.html.

23. This term is frequently used in explaining the need for national environmental laws and uniform environmental standards. For an early examination of this argument, see Mary Graham, "Environmental Protection and the States: Race to the

Bottom or Race to the Bottom Line?" (Washington, DC: Brookings Institution, December 1, 1998), www.brookings.edu/articles/environmental-protection-the-states-race-to-the-bottom-or-race-to-the-bottom-line/ (accessed September 24, 2017).

24. Michael Gorn, "William D. Ruckelshaus: Oral History Interview."

25. Interview with William Ruckelshaus for "Poisoned Waters" (transcript), *Frontline,* April 21, 2009, www.pbs.org/wgbh/pages/frontline/poisonedwaters/interviews/ruckelshaus.html (accessed July 9, 2015).

26. Ibid.

27. J. Gordon Edwards, "DDT: A Case Study in Scientific Fraud," *Journal of American Physicians and Surgeons* 9, no. 3 (2004): 83–88, 86.

28. David Kinkela, *DDT and the American Century: Global Health, Environmental Politics, and the Pesticide That Changed the World* (Chapel Hill: University of North Carolina Press, 2011), 159.

29. EPA, "DDT Regulatory History: A Brief Survey (to 1975)," excerpt from "DDT: A Review of Scientific and Economic Aspects of the Decision to Ban Its Use as a Pesticide," prepared for the House Committee on Appropriations, July 1975, EPA-540/1-75-022, https://archive.epa.gov/epa/aboutepa/ddt-regulatory-history-brief-survey-1975.html.

30. See, for example, Denise Scheberle, *Environmental Federalism: Trust and the Politics of Implementation* (Washington, DC: Georgetown Press, 2004), 1–26.

31. See Gorn, "William D. Ruckelshaus: Oral History Interview," for a description of President Nixon's interactions with the EPA.

32. David Gutman, "Like Sally Yates, William Ruckelshaus Said 'No' to a President—and Got Fired," *Seattle Times,* January 30, 2017.

33. Andrew Ramonas, "Remembering Watergate: Ruckelshaus Gives Riveting Account of Saturday Night Massacre," includes text of a speech given by William Ruckelshaus to former US attorneys, October 3, 2009, www.mainjustice.com/2009/10/08/ex-nixon-official-gives-riveting-account-of-saturday-night-massacre/ (accessed July 10, 2015).

34. "Top 10 Best Cabinet Members," *Time,* 2009, http://ti.me/1fjOIeH (accessed June 24, 2017).

35. Patricia Sullivan, "Anne Gorsuch Burford, 62, Dies; Reagan EPA Director," *Washington Post,* July 22, 2004.

36. Phil Wisman, "EPA History (1970–1985)" (Washington, DC: EPA, November 1985), https://archive.epa.gov/epa/aboutepa/epa-history-1970-1985.html.

37. Joel Dyer, "How Reagan and the Largest EPA Scandal in History May Explain Why Valmont Butte Is Still Contaminated," *Boulder Weekly,* February 9, 2012; see also Steven R. Weisman, "President Names Ruckelshaus Head of Troubled EPA," *New York Times,* March 22, 1983.

38. Ralph Frammolino, "Lavelle Serving Sentence in San Diego: Convicted EPA Official Speaks Out on 'Plot' Against Her," *Los Angeles Times,* May 11, 1986.

39. Gorn, "William D. Ruckelshaus: Oral History Interview."

40. "Environmental Protection, Paralyzed" (editorial), *New York Times*, November 30, 1984.

41. William Ruckelshaus, "Choosing Our Common Future: Democracy's True Test," Fifth Annual John H. Chaffee Memorial Lecture, National Council for Science and the Environment, Washington, DC, February 3, 2005, 14.

42. Ibid., 24.

43. Sudjit Luanpitpong, Michael Chen, Travis Knuckles, Sijin Wen, Juhua Luo, Emily Ellis, Michael Hendryx, and Yon Rojanasakul, "Appalachian Mountaintop Mining Particulate Matter Induces Neoplastic Transformation of Human Bronchial Epithelial Cells and Promotes Tumor Formation," *Environmental Science and Technology* 48 (October 2014): 12912–12919, dx.doi.org/10.1021/es504263u.

44. Michelle Nijhuis, "West Virginia Activist Julia Bonds Takes on Mountaintop Removal Mining," *Grist*, April 15, 2003, http://grist.org/article/slaughter/ (accessed on March 17, 2016).

45. Nicole Makris, "A Coal Miner's Daughter," *Mother Jones*, June 16, 2003, www.motherjones.com/environment/2003/06/coal-miners-daughter (accessed March 17, 2016).

46. Nijhuis, "West Virginia Activist Julia Bonds Takes on Mountaintop Removal Mining."

47. Ibid.

48. Taylor Lee Kirkland, "Mountain Memories: Interview with Judy Bonds," *Yes*, January 7, 2011. http://yesmagazine.org/people-power/interview-with-judy-bonds (accessed March 22, 2016).

49. Quoted in Jeff Biggers, "Thousands Pay Tribute to Judy Bonds: She Has Been to the Mountaintop—and We Must Fight Harder to Save It," *Huffington Post*, May 25, 2011, www.huffingtonpost.com/jeff-biggers/thousands-pay-tribute-to_b_804001.html (accessed February 9, 2016).

50. Goldman Environmental Foundation, "Julia Bonds, 2003 Goldman Prize Recipient," 2003, www.goldmanprize.org/recipient/julia-bonds/ (accessed February 9, 2016).

51. Nijhuis, "West Virginia Activist Julia Bonds Takes on Mountaintop Removal Mining."

52. Ibid.

53. Coal River Mountain Watch, "About Us," www.crmw.net/about.php (accessed February 2, 2015).

54. Dennis Hevesi, "Judy Bonds, an Enemy of Mountaintop Coal Mining, Dies at 58," *New York Times*, January 15, 2011.

55. Nijhuis, "West Virginia Activist Julia Bonds Takes on Mountaintop Removal Mining."

56. Goldman Environmental Foundation, "Julia Bonds: 2003 Goldman Prize Recipient."

57. Ibid.

58. Ibid.

59. Nijhuis, "West Virginia Activist Julia Bonds Takes on Mountaintop Removal Mining."

60. Emma Brown, "Miner's Daughter Fought Mountaintop Removal," *Washington Post,* January 6, 2011.

61. Makris, "A Coal Miner's Daughter."

62. Elizabeth Shogren, "Bush Administration Altered Appalachian Landscape," *National Public Radio, Weekend Edition,* January 17, 2009, www.npr.org/templates /transcript/transcript.php?storyId=99493874 (accessed March 31, 2016).

63. Ken Ward Jr., "Bush Administration Plan Broadens Valley Fill Rule Changes," *Charleston Gazette-Mail,* April 26, 2002.

64. Appalachian Voices, "Stream Protection Rule," http://appvoices.org/end -mountaintop-removal/stream-protection-rule/ (accessed March 31, 2016).

65. Noelle Straub and Eric Bontrager, "Salazar Moves to Overturn Bush Admin Mountaintop Rule," *E&E News PM,* April 30, 2009, www.eenews.net/landletter /stories/77377/ (accessed March 15, 2016).

66. Jeff Biggers, "Nonviolent Goldman Prize Winner Attacked by Massey Supporter: 94-Year-Old Hechler, Hannah, Hansen Arrested at Coal River," *Huffington Post,* Green Blog, July 24, 2009, www.huffingtonpost.com/jeff-biggers/live-at-coal -river-daryl_b_219628.html (accessed May 12, 2016).

67. Eric Bontrager, "Climate Scientist Hansen, Actress Daryl Hannah Arrested During Mountaintop Protest," *E&E News PM,* June 23, 2009, www.eenews.net /eenewspm/stories/79575/ (accessed March 15, 2016).

68. Eric Bontrager and Taryn Luntz, "Few Cheers for the Obama Admin's Mountaintop Plan," *E&E News PM,* June 11, 2009, www.eenews.net/eenewspm/stories /79129/ (accessed March 15, 2016).

69. Earthjustice, "Federal Court Strikes Down Bush-Era Stream-Dumping Rule," February 20, 2014, http://earthjustice.org/news/press/2014/federal-court-strikes -down-bush-era-stream-dumping-rule (accessed March 31, 2016).

70. Devin Henry, "Trump Signs Bill Undoing Obama Coal Mining Rule," *The Hill,* February 16, 2017, http://thehill.com/policy/energy-environment/319938 -trump-signs-bill-undoing-obama-coal-mining-rule (accessed September 24, 2017).

71. Robert Shetterly, "Judy Bonds: Environmental Activist, 1952–2011," Americans Who Tell the Truth: Models of Courageous Citizenship, www .americanswhotellthetruth.org/portraits/judy-bonds (accessed May 12, 2016).

72. Coal River Mountain Watch, "Judy Bonds at PowerShift07!" 2007, YouTube, https://youtu.be/XjeFj2NjdQY (accessed May 12, 2016).

73. "Capitol Climate Action: Mass Civil Disobedience in DC Against Use of Coal at Capitol Hill Power Plant" (transcript), *Democracy Now,* March 2, 2009, www .democracynow.org/2009/3/2/capitol_climate_action_thousands_converging_on (accessed May 12, 2016).

74. Tim Huber, "Julia Bonds, 58, Mountaintop Mining Opponent," *Boston Globe,* January 7, 2011.

75. "Coalminer's Daughter Turned Activist Wins Top Enviro Prize" (transcript of interview), *Living on Earth, Public Radio International,* April 18, 2003, www .loe.org/shows/segments.html?programID=03-P13–00016&segmentID=7 (accessed June 24, 2017).

76. Jason Howard, "Appalachia's Patron Saint," The HillVille, January 5, 2012, http://thehillville.com/2012/01/05/appalachias-patron-saint/ (accessed March 22, 2016).

77. Vernon Haltom, "You Are the Ones You've Been Waiting For!" *Coal River Mountain Watch Messenger* 5, no. 1 (2011): 1–2.

CHAPTER 7

The Rest of Us

LEARNING HOW TO BE ENVIRONMENTAL HEROES

The final chapter of the book belongs to "the rest of us." Throughout the book, you may have booed the villains who brought about the loss of tens of thousands of lives in Bhopal, India, the continuing loss of life in Libby, Montana, the biggest oil spill in US history from the blowout at the Macondo well, and the catastrophic Upper Big Branch Mine explosion. At the same time, you may have cheered the heroes in these stories who sounded the alarm about the dangers, as well as the heroes highlighted in the previous chapter: Gaylord Nelson, William Ruckelshaus, and Judy Bonds. Now it is time to consider how the rest of us can shape environmental stories.

This chapter begins by looking at the lessons we might draw from these environmental stories—like learning to delineate between "normal" accidents and those that are the result of irresponsible company culture and ineffective government oversight, or understanding how highly reliable organizations come to be. By drawing from these lessons, we may be able to minimize tragedies like the ones described in Chapters 2 through 5. Next, the chapter focuses on ways in which organizations can become more environmentally focused and move beyond high reliability to become greener, more sustainable organizations.

The chapter ends, appropriately, with all of us. How do we become the everyday environmental heroes that the planet needs? Environmental problems are "wicked problems" in that they are societal problems not easily solved.[1] They are problems that involve different values and a wide array of stakeholders. Climate change is a good example, as it requires a new way of thinking and acting in order to avoid a devastating warming of the planet.[2] Our one blue planet depends on us acting in both individual and collective ways, in our communities and our organizations, to make it a better place. The chapter concludes with reflections on how we can indeed change the world.

ENVIRONMENTAL STORIES: LESSONS LEARNED

What lessons should we draw from our environmental stories? First, the stories of Union Carbide, BP, Massey Energy, and W. R. Grace help illuminate the difference between "normal" accidents—ones that are inevitable because of complex and tightly coupled systems—and the disasters that occurred because of the neglectful, or even villainous, conduct of key personnel in these organizations. Second, these stories demonstrate that government can respond to the magnitude of these disasters and the intense media attention and public criticism with policy shifts and regulation changes.

Distinguishing Between Normal Accidents and Disasters Due to Organizations Run to Failure

The strong, safety-minded cultures found in highly reliable organizations encourage everyone in the organization to be vigilant and responsive to problems as soon as they develop. Industrial disasters like those profiled in Chapters 2 to 5, and even "normal" accidents, are thus less likely to occur in such organizations. Members of highly reliable organizations undertake their tasks with a "mindfulness" and a healthy disregard for formal hierarchy that suggest a culture in which problems are not just passed up the chain of command, but that cautions everyone to be aware of potential failures in the system.[3] Mindfulness helps organizations strive for and maintain best practices—behavior that some organizations are able to sustain for a long period of time.

Compare this kind of thinking to the mind-set in less reliable organizations, like those discussed in this book. Vigilance may atrophy in these organizations as workers become accustomed to paying attention to the organization's outputs or products and less likely to notice or fix potentially hazardous conditions. Instead of watching for small anomalies or risk factors, employees become complacent. As complacency sets in, organizational standards are gradually eroded and deviations from safety protocols become normalized.[4] As a result, the organization becomes less capable of taking ordinary precautions to minimize risk. Instead of practicing mindfulness, managers and staff assume that because nothing has gone wrong before when safety processes are ignored, nothing ever will. Atrophy of vigilance and normalization of deviance lead to a kind of *mindlessness* in an organization—just the opposite of what high-reliability organizations strive for.

Indeed, the stories recounted here show that safety was not a priority for the leaders in these organizations, who were more likely to put profits first, with an eye toward minimizing costs. For example, Massey Energy executives encouraged mine managers to change operations to accommodate inspections of the mine, only to return to unsafe ventilation and coal dust handling methods once the federal or state inspector left the premises. You'll recall from Chapter 4 that investigators cited the "normalization of deviance" at Massey as a factor in the Upper Big Branch explosion. That accident was not surprising; the surprise was that it had not happened sooner, given Massey's lax attention to safety procedures.

Similar conclusions can be drawn from the Bhopal story in Chapter 2. Like the Upper Big Branch mine, the Union Carbide plant was an accident waiting to happen. There were widespread problems at the plant, from design flaws and operating errors to relaxation of maintenance. Workers were poorly trained, and previous leaks in the storage tanks had been largely discounted. Refrigeration designed to keep methyl isocyanate cool had been shut down in anticipation of the plant's closure. Workers were ill prepared to handle emergencies, and in any event, fewer workers were present to handle them because Union Carbide had cut the staff from twelve to six operators.[5] Training levels had been similarly reduced. And Union Carbide executives in India, looking to avoid the panic caused by previous accidents, decided to turn off the alarm system's warning bell, which would have alerted thousands of Bhopal residents of the danger during the "Night of the Gas."

As Charles Perrow would observe in the second edition of his book *Normal Accidents,* Bhopal's tragedy was the result not of system failure, but of greed on the part of company executives, as well as social and economic conditions. Perrow notes that the extensiveness of the catastrophe—thousands of people killed in the first few days, with hundreds of thousands affected in the following months and years—resulted from the presence of high quantities of highly toxic chemicals, the lack of a warning system, the proximity of large numbers of people who were unprotected and unaware of the danger, and a lack of emergency response once the accident occurred.[6] For Perrow, the critical element needed to prevent stories like Bhopal from ever being told again is to reduce the catastrophic *potential:* "An economic system that runs such risks for the sake of national prestige, patronage, or personal power, is the more important focus and culprit . . . the issue is not risk, but power."[7]

So, too, the BP oil spill in the Gulf of Mexico is less the story of a "normal" or systems accident than one of a corporate culture that looked for decades to cut costs in its desire to be one of the world's leading oil producers. After the Texas City explosion in 2005, investigations discovered a number of issues at the BP refinery, including what the staff would describe as the organization's policy of "running to failure" rather than scheduling maintenance on equipment.[8] Workers warned, just as the miner Gary Quarles did about conditions at the Upper Big Branch mine, that a "culture of casual compliance" existed at the refinery and that they were exceptionally fearful that a catastrophic incident could occur at Texas City.[9]

When the *Deepwater Horizon* exploded five years later, BP would claim that the accident was the result of a series of unexpected events, but federal investigators would find instead that BP shortcuts were largely to blame. In the face of cost overruns on the *Deepwater Horizon,* BP was eager to seal the Macondo well and move on. The report would note:

> The loss of life at the Macondo site on April 20, 2010, and the subsequent pollution of the Gulf of Mexico through the summer of 2010 were the result of poor risk management, last-minute changes to plans, failure to observe and respond to critical indicators, inadequate well control response and insufficient emergency bridge response training by companies and individuals responsible for drilling at the Macondo well and for the operation of the *Deepwater Horizon.*[10]

Perhaps the starkest example of the normalization of deviance in the four stories was how W. R. Grace operated its vermiculite mine in Libby, Montana. Company executives knew for years that the asbestos-containing vermiculite caused cancer, but allowed respirators to go unused by the mine workers and conditions to continue at the dry mill. Worse, evidence presented at trial documented that company executives suppressed the information that asbestos from the mine was also poisoning families and threatening the community. The result was a town with a rate of lung disease forty to sixty times the national average. The US government has described the mine operation in Libby as "the worst case of industrial poisoning of a whole community in American history."[11]

In each of these cases, conduct on the part of many key people in the organization was negligent at best, and villainous at worst. It is hard to excuse these industrial disasters as "normal" accidents, and it is equally impossible to label any of the organizations responsible for them as "highly reliable" at the time.

Public Policy Responses to Disasters: Agenda-Setting and Punctuated Equilibrium

If one lesson we can draw is that we must draw a sharp line between normal accidents and the disasters recounted here, another lesson relates to changes in public policy. As described in Chapter 1, items are often moved onto a governmental agenda in response to a focusing event that captures media and public attention. These four tragedies triggered such an agenda-setting response, though the extent and timing of these responses varied.

Additionally, long-neglected policies that have been changed only incrementally for years can sometimes be changed dramatically in response to perceived defects in existing practice. These large-scale departures from policy stability, as explained by "punctuated equilibrium" theory, occur when agenda-setting issues prompt policymakers to rethink how certain public problems have been addressed. Although governmental policies concerning the environment and corporate safety and work standards were in place before these four disasters occurred, new approaches to regulating toxins, offshore drilling, underground mining, and asbestos followed in their aftermath.

Without a doubt, the BP oil spill was front-page news, and the live feed of oil spewing from the bottom of the Gulf was a constant reminder of the ongoing disaster. In response to the public outcry, the Obama administration quickly launched an investigation. What it found was embarrassing not only to BP but also to the federal government. The discovery of the Minerals Management Service's (MMS's) regulatory failures prompted its reorganization into separate agencies for revenue collection, resource management, and enforcement duties. Those changes happened quickly: in just over a month, Secretary of the Interior Ken Salazar signed the reorganization order, and the Office of Natural Resources Revenue, charged with collecting revenues from offshore drilling, became a separate entity, on October 1, 2010. Two new, independent bureaus, the Bureau of Safety and Environmental Enforcement (BSEE) and the Bureau of Ocean Energy Management (BOEM), were operational on October 1, 2011.[12]

The newly created Bureau of Safety and Environmental Enforcement (BSEE) was charged with vigorously enforcing offshore safety and environmental regulations, and the inspection and engineering workforce nearly doubled, from fifty-five in 2010 to ninety-two in 2016.[13] The agency acted to reduce risk through better well design and casing standards and to promulgate the "Safety and Environmental Management System Rule," which in part empowered workers to participate in safety management decisions. Audits conducted by accredited third parties would now be required. New safeguards took effect that would protect offshore personnel and the environment, such as the drilling safety rule finalized in 2012.

However, not all changes were simple. Especially contentious was a rule that focused on the blowout preventer, referred to as the "Well Control Rule." The rule required that inspectors be on location, that they observe testing of the blowout preventer prior to any drilling, and that the best technology be used. Setting this standard proved a hard slog for the Obama administration, as oil and gas interests strongly opposed any new regulations on the grounds that they were too costly and difficult to implement. The final regulation would not be finalized until six years after the BP spill. The fight over future drilling sites quickly became a political minefield pitting fossil fuel proponents against environmental groups.

Thus, a dramatic event—the BP oil spill—focused the attention of the president, Congress, and the executive branch on the need to reform the regulatory system and on the dangers of offshore drilling. Environmental regulations were changed, as were the procedures for enforcing them. Moreover, an agency was abolished and replaced by three others with single-purpose missions.

However, the oil industry was not stymied for long by these changes. Although the Obama administration placed a moratorium on offshore drilling, it would be short-lived as people in the Gulf region pressed for renewal of the economic benefits of offshore drilling. The moratorium was lifted in October 2010, and permits to drill in deep water gradually increased in number, much to the chagrin of environmental groups. In the waning days of his presidency, Obama banned all future offshore oil and gas drilling from nearly 120 million acres of land in the Atlantic and Arctic Oceans. However, the Trump administration is taking the opposite approach by promoting offshore drilling and opening up the Arctic and the Arctic National Wildlife Refuge to new drilling.

Agenda-setting and policy changes were also evident in the other three stories. After Bhopal, Congress passed the Emergency Planning and

Community Right-to-Know Act (EPCRA) in 1986 in response to concerns about the environmental and safety hazards posed by the manufacture, storage, and handling of toxic chemicals. Congress sought to prevent a similar disaster in American communities by arming them with information—organizations had to report on-site chemicals, how they were being processed, and any releases into the environment. Over time the Toxics Release Inventory (TRI), the compilation of information about toxins at facilities, became a useful gauge of potential risks due to chemical releases.

In a good example of punctuated equilibrium, the aftermath of the explosion at the Upper Big Branch mine prompted the Mine Safety and Health Administration (MSHA) to undertake a comprehensive internal review. As a result of that review, MSHA took over one hundred corrective actions and has reported updates of its progress on its website. Some of these corrective actions included enhanced enforcement programs; reorganization of the Office of Assessments, Accountability, Special Enforcement and Investigations in order to better manage enforcement; additional training; and revision of forty policy directives, including procedures for mine inspections.[14] Equally important, whistleblower protections were put in place, inspections were made more thorough, and safety violations are now more closely tracked to identify repeat offenders.

These are laudable efforts, but they were made only after the MSHA felt the ire of the public. The agency should have shut the mine before the explosion, given the pattern of serious safety violations, but made no attempt to do so. In fact, MSHA had not shut a mine for thirty-four years.[15] After the disaster at the Upper Big Branch mine, the MSHA revised its rule on closing mines and used it in 2011—which would be the safest year in MSHA history—to initiate shutdown procedures in Kentucky and West Virginia.

Recently, however, the Trump administration has initiated friendlier relations with the coal industry. Vice President Pence declared during the 2016 debates that he would stop the "war on coal" caused by new environmental regulations by moving to relax those regulations and bring a fossil-fuel-friendly administration to the White House. The Trump administration and the Congress have made good on these campaign promises, reversing some of the Obama administration's coal rules, such as the stream protection rule described in Chapter 6. In truth, the decline in coal jobs has been due to technological advances, including longwall and mountaintop removal mining, and the demand for coal has declined as power plants shift to natural gas and renewable energy, putting a double whammy on any attempt to revive the industry.

The W. R. Grace story unfolded as the country was reshaping its policy toward asbestos. Asbestos class action suits were under way, and public concern was reinforced by the poisoning of Libby, Montana, where the EPA declared the first-ever public health emergency, noting the severity of the risk from this carcinogen. The same year the EPA placed Libby on the Superfund program's National Priorities List, Senator Patty Murray (D-WA) introduced the Ban Asbestos in America Act. It would finally be passed in the Senate in 2007, but not in the House of Representatives.[16] Fearing that the Libby story might unfold at other mines, the EPA inspected vermiculite mines, and samples were tested for possible asbestos contamination. Initial reports did not reveal asbestos fibers. Left unanswered was the question of how much people whose homes were fitted with insulation made from Zonolite vermiculite had been exposed to asbestos.

These stories also followed the issue-attention cycle described by Anthony Downs (see Chapter 1).[17] Each story captured public attention and resulted in public demand for action, though to different degrees. Without a doubt, the BP oil spill occupied center stage for the longest time—in part because the spill was massive, but also because the live feed of oil gushing from the Gulf floor made for mesmerizing television coverage. The Bhopal explosion had the greatest human impact: half a million people have been affected by that disaster. Although the tragedy was front-page news in the American media for a time, its legacy effects, including serious environmental contamination around the closed Union Carbide plant, have remained largely off the radar of the American public. The stories of all four disasters and the resulting policy responses have largely moved off of the formal agenda as public attention to the issues that emerged during these disasters has waned. Those issues now largely occupy the post-problem stage, a part of the issue-attention cycle described by Downs as a "twilight realm of lesser attention," and they will languish there until either a similar disaster occurs or a new problem opens another chapter of these stories.[18]

A BETTER WAY: MOVING TOWARD SUSTAINABLE AND ACCOUNTABLE ORGANIZATIONS

The four disasters recounted in this book prompted new laws and regulations and changes in regulatory agencies that in the future will help monitor other organizations that may be similarly inclined to act with little regard for health, safety, and environmental risks. These disasters also proved costly to each business: their stock prices plummeted, Union Carbide and Massey Energy were acquired by larger companies, W. R. Grace filed for bankruptcy, and BP paid more in fines and environmental cleanup costs than any company in history. The most important takeaway is that we can expect much better performance by organizations. Not only can catastrophes be minimized, but organizations can become environmental leaders, and everyday citizens can help hold them accountable.

Corporations: Building Sustainability and the Triple Bottom Line

Corporations are increasingly realizing the benefits of "going green." Some have taken on the mantle of ecological stewardship by going beyond measuring company performance by just profits and returns on investments and also including environmental and social dimensions. John Elkington, who pioneered this kind of accounting in the 1990s, refers to it as the Triple Bottom Line (TBL).[19] The TBL components are often called "the Three Ps"—people, planet, and profits. A closely related concept, sustainability, is often defined as the balance of "the Three Es": equity, environment, and economy. Some of the companies that have eschewed the model of profit at any cost have proven that a company can be sustainable and highly reliable while also staying at the top of its industry.

For example, in 2016, Google, Amazon, Apple, and Microsoft filed a legal brief supporting the EPA's Clean Power Plan.[20] Both international companies and nongovernmental organizations (NGOs) participated in commitments to shift to renewables as part of the Paris Agreement. Nearly half of the Fortune 500 companies in the United States have set targets to reduce their carbon footprints, and nearly two dozen have pledged to power their operations with 100 percent renewable energy.[21] As David Wei and his colleagues observed, "For the first time the private sector is recognized as an integral part of the global solution to address climate change. There is a clear policy signal for businesses and investors across all jurisdictions to make low emission or emission-neutral investments, whether through financing projects or investing in new technologies."[22] Moreover, the greening of business is not limited to the United States. More than eight hundred of the world's largest companies supported a global agreement to tackle climate change.[23]

Although this book has largely focused on the private sector, it is worth noting that governments and NGOs may also suffer from normalization of deviance and bad practices and thus could also be improved by cultivating high-reliability cultures and applying TBL principles. Many local governments and nonprofit organizations have adopted sustainability principles and "greener" practices. These governmental sectors obviously have no profit incentive for their actions, but they do seek economic prosperity, social well-being, and environmental protection.[24]

The Rest of Us: Holding Corporations and the Government Accountable

As we learned from the disaster stories, government policies and regulations can change in the aftermath in response to the media attention and public scrutiny brought on by the disaster. And as we have seen from the earlier discussion here, some corporations and other organizations have become more environmentally conscious and adopted more ecologically sound practices—in part because of pressure from their customers or from citizens. However, even though the disasters described in this book and the positive changes they prompted happened a while ago, and even though some corporations now see the benefits of becoming more sustainable, there are still some companies and government organizations that need to be persuaded to do the right thing.

That's where the rest of us come in. There are many things we can do to hold organizations—both corporations and the government—accountable in the hopes that they will become more environmentally conscious, more responsible, and more transparent about their practices. Most companies would like to be recognized for taking corporate responsibility seriously, and we can encourage them.

For example, we can insist that companies do business in an environmentally sensitive manner, and we can support them when they do so by buying their products. A well-known example is the consumer boycott of tuna organized by Earth Island Institute's International Marine Mammal Project and the Sea Shepard Society in 1988. Over the next two years, the grassroots campaign compelled the three largest tuna companies in the world to adopt tuna-fishing practices that avoided the netting of dolphins. Today 90 percent of the world's canned tuna markets demand "dolphin-safe" tuna-fishing methods.[25]

We can also encourage companies to "go green" by monitoring how they use energy, produce goods, and treat employees and stakeholders. We can look for green companies in the news or review lists of companies that have adopted sustainable business practices. *Newsweek,* for example, publishes an annual ranking of the five hundred largest US businesses based on the companies' environmental impact, environmental management, and environmental disclosure. Similarly, a group called the Corporate Knights annually ranks the top one hundred most sustainable corporations in the world and releases that ranking through the World Economic Forum.[26]

Subsequent coverage by business news organizations such as *Forbes* makes the information accessible to us and lets companies know that their green efforts are newsworthy.[27] On the flip side, organizations such as the Center for Public Integrity identify the worst-of-the worst polluting companies. Called "America's Super Polluters," that list provides the kind of negative publicity no company wants.[28]

We also can educate ourselves about companies and organizations through government documents and reports. EPCRA, the law passed by Congress after the Bhopal tragedy, requires companies to submit reports on their toxic chemical releases to the EPA. The EPA compiles this information into the TRI, described previously, which tells us which chemicals are being released into our air and water or deposited on the land. We can learn more about local companies by accessing the TRI database on the EPA website. Indeed, the EPA and state environmental agencies provide several searchable databases to help make us aware of what is happening in our communities.

If a company is publicly traded, we can examine the annual report to stockholders to see what commitments have been made to shift to greener ways of doing business or what local initiatives the company supports. As stockholders or consumers, we can continue to push for disclosure so that we can assess the progress (or lack thereof) made by companies. We can contact companies directly too, to let them know that we are monitoring their sustainability efforts. As noted later in the chapter, we can also pressure organizations that own stocks of companies with poor performance records to sell the shares they own. These are some of the ways in which our voices can be heard in the decision-making structure of a company.

Another powerful tool for "the rest of us" is the citizen suit provision found in many environmental laws. A citizen suit provision allows us to steer the course of environmental protection efforts in federal and state agencies by requiring that these agencies enforce regulatory requirements. Government agencies are subject to budget cuts and varying levels of political support. They need to be monitored too, as suggested by the water supply problem in Flint, Michigan (discussed later in the chapter), as well as in our previous stories. Citizen suit provisions allow citizens to sue an agency for failing to perform a nondiscretionary duty, such as promulgating new environmental standards or issuing violations to recalcitrant companies. These provisions, along with the ability to comment on changes in agency policies, help us hold our public organizations accountable.

Citizen suits also provide citizens with an opportunity to act as private attorneys general to bring lawsuits against companies that violate environmental standards. When writing national environmental laws, Congress anticipated that there would be enforcement gaps in their implementation. Thus, citizens, typically represented by environmental groups, can compel the inspection and enforcement activities of agencies, most notably the EPA, but they can also be "deputized" to directly sue companies that violate environmental laws. In short, if regulatory agencies are not willing or able to enforce environmental standards, we can.

As we have seen from the Bhopal explosion story in Chapter 2, the Libby, Montana, asbestos story in Chapter 5, and the stories of environmental heroes in Chapter 6, journalists and citizens have played a big role in bringing the shady practices of corporations to light and holding them accountable for their misdeeds. This is evident in more recent events as well, like the ongoing disaster in Flint, Michigan. In 2014, hoping to save money, the state-appointed emergency manager for the city of Flint switched its water supply source from the Detroit Water and Sewerage District to the Flint River. However, the switch revealed that drinking water from the river was contaminated with bacteria. After issuing two boil-water advisories due to fecal and total coliform bacteria levels, the city of Flint increased the amount of chlorine it used to decontaminate the drinking water supplies, and it flushed the system. But now the city had to warn residents of unsafe levels of cancer-causing disinfectants in the drinking water.[29] At the same time, the Flint River was naturally high in corrosive chloride. In a dramatic example of the complexity of public problems, when the city switched to the river for its water source, it opted not to add a corrosion inhibitor to its water treatment in order to save about $140 per day.[30]

In 2015, the EPA found another frightening contaminant in Flint's drinking water: lead. Flushing and decontaminating the aging and inadequate pipes that brought water into Flint homes had caused dangerously high levels of lead to leach into the water, prompting a public health emergency. Lead exposure at high levels is dangerous, especially for children, in whom it can impair cognitive function and cause behavioral disorders and hearing problems. Lead can also affect the heart, kidneys, and nerves.

The city began corrosive control treatment of the water supply, but it was too little too late. In 2016, Governor Rick Snyder declared a state of emergency and mobilized the Michigan National Guard to distribute clean water. Two weeks later, President Obama authorized federal assistance and declared a state of emergency in Flint. State and local officials had failed to promptly warn Flint residents about the dangerous levels of lead in their drinking water. Criminal charges were filed against government employees, along with lawsuits under the Safe Drinking Water Act and class action lawsuits from residents affected by the lead contamination. The final bill for installing new pipes and fixing Flint's water supply system could run as high as $1.5 billion; the eventual damages awarded under pending lawsuits could add many millions more to the cost.[31]

Why did residents' exposure to lead in the water persist for so long? It seems obvious that our government agencies should have declared an emergency as soon as they knew that the water was contaminated. Instead, just as Gayla Benefield and Judy Bonds had done, it was citizens who sounded the alarm. LeeAnne Walters, a mother of four and leader of Water Warriors, an impromptu citizens' group, took action after her three-year-old twins developed rashes and showed symptoms of ill health.[32] She began going to city council meetings, voicing her concern along with other citizens. She listened to Flint's mayor, Dayne Walling, declare that the city water was safe to drink, but she refused to back down. Finally, she persuaded the city to test her water. The results were frightening: the lead concentration in the water from her tap was 400 parts per billion (ppb). The maximum concentration permitted under the Safe Drinking Water Act is 15 ppb.

When she discovered the extent of lead contamination in Flint's water, Walters scoured the city's reports required under the Safe Drinking Water Act and learned that the water from the Flint River was more corrosive than the water that had been previously supplied by the Detroit Water and Sewage District. She also learned that the testing methods employed by Flint officials were designed to underreport the extent of the contamination. City officials allowed the water to run a few minutes before sampling, flushing the pipes of the worst of the lead contaminants.

Walters's delivery of what she had found to the regional EPA triggered a new round of tests in her home. This time proper testing revealed a staggering 13,200 ppb of lead coming from her tap.[33] Had she and other concerned citizens not acted, one wonders how long their lead exposure might have persisted.

LeeAnne Walters had the tenacity and courage to be an environmental hero. The next section looks at a problem that offers all of us the opportunity to act heroically: climate change.

BECOMING AN EVERYDAY ENVIRONMENTAL HERO: ADDRESSING CLIMATE CHANGE

Curbing the impact of climate change is one area where "the rest of us" can, and should, become involved. Most scientists consider climate change to be the greatest environmental foe of this century. Stopping the warming of a planet seems like a daunting endeavor, and at first glance, it seems impossible that we could bring about positive change. For several reasons, however, considering the actions we can take to address global warming illustrates how we can become everyday environmental heroes.

First, climate change is arguably the most challenging of all environmental problems. Second, tackling this problem requires cooperation and heroic action from everyone—governments, corporations, nonprofit organizations, and individual citizens. Moreover, climate change affects the entire global village—no country will escape the consequences of global warming. Finally, as challenging as it is to address climate change, its worst consequences are not inevitable. We have it in our power to slow the warming of our planet. However, the longer we delay, the more unlikely it is that we will avoid the most damaging effects of climate change. To paraphrase Judy Bonds, we need to act as though our continued existence depends on curbing climate change—because it just might.

Climate change is already having significant impacts. Global surface temperatures in 2016 were the warmest since record-keeping began in 1880, according to analyses by the National Aeronautics and Space Administration (NASA) and the National Oceanic and Atmospheric Administration (NOAA).[34] In 2016, May temperatures soared to their hottest in 137 years, with average increased temperatures of 1.57 degrees Fahrenheit, and 2016 was the hottest year on record.[35] But 2017 was equally problematic. In 2017, planes were grounded in Phoenix, Arizona, and Las Vegas, Nevada, as temperatures reached 121 degrees (planes cannot take off when temperatures exceed 118 degrees), and a blistering heat wave along the West Coast toppled record highs.

People around the globe suffered in 2016 and 2017 from heat-related illnesses, even death. India experienced its highest temperature ever recorded at 123.8 degrees Fahrenheit on May 20, 2016, and the intense heat brought misery for millions of people who had little access to water and no access to electricity. Hundreds succumbed to heat-related illness and death.[36] This followed a death toll of over two thousand from the previous year's heat wave. But it wasn't just India. Ahvaz, Iran, reached 129 degrees in June 2017—the highest temperature ever recorded for that country. NOAA found that January to May 2016 temperatures across global land and surface areas was an alarming 1.94 degrees Fahrenheit above the twentieth-century average—making this time period the hottest since record-keeping began in 1880.[37] Sadly, climate scientists predict these temperature increases will continue, and perhaps accelerate.

The year 2017 saw a hurricane season that the United States has never experienced. Harvey, Irma, and Maria made landfall in the United States or its territories, causing loss of life, catastrophic flooding, and a wide swath of property damage. When Hurricane Harvey made landfall in August, an incomprehensible 19 trillion gallons of rain fell in Texas, with an estimated recovery price tag of $180 billion.[38] Hurricane Irma made landfall in Florida on September 10, 2017, knocking out power to millions of people; Hurricane Maria hit Puerto Rico and the Virgin Islands just ten days later, devastating the islands and leaving Puerto Rico's three million residents without power. Most scientists link a hotter planet, with rising ocean temperatures, to increased extreme weather events.

The presence of feedback loops in the climate system makes time a critical factor. A feedback loop acts like a vicious circle: as its elements intensify and aggravate each other, it may spiral out of control. In the case of climate change, a positive feedback loop accelerates the pace of temperature rise. One example of a positive feedback loop in climate change is melting ice. Ongoing global heat has accelerated the melting of the Arctic sea ice, bringing ice coverage to the lowest levels ever documented. Ice helps keep the planet cool by reflecting sunlight back into space. Land ice sheets in Antarctica and Greenland have also been losing mass, according to NASA satellite data.[39] With less ice, the planet absorbs more of the sun's energy, which leads to more global warming, which in turn leads to more melting ice, and so on. Also, as land ice melts sea levels rise, threatening coastal areas and increasing the likelihood of flooding.

Increasing levels of greenhouse gases, such as carbon dioxide and methane, have also driven the extreme weather events shattering records across the planet. Even Antarctica has not escaped the effects of climate change. In June 2016, carbon dioxide concentrations reached 400 parts per million (ppm), a concentration of greenhouse gas not present for four million years. Scientists have long viewed 350 ppm as the upper limit on "safe" levels of CO_2 in the atmosphere.[40] CO_2 levels are now past 406.69 ppm (as of July 2017) and steadily rising, according to data compiled by NASA, reminding us that we risk triggering irreversible impacts that could send climate change spinning out of control. As *Scientific American* observed, "We're officially living in a new world."[41]

With the advance of climate change, scholars have called for nothing less than an ecological citizenship that will reshape individual and community identities, invigorating a people-power energy to meet the challenge of a rapidly changing planet. Ecological citizenship expands the concept of environmental citizenship, in that it sees our environmental responsibilities as transcending any territorial border. Instead of putting nation-states solely in charge, ecological citizenship locates responsibility in each of us to reduce our individual ecological footprint and the demands we place on nature. Websites such as the Global Footprint Network provide tools that enable us to see how much land area is needed to support our individual lifestyles in the hope that, as ecological citizens, we can learn to tread more lightly on the earth.[42] Those who find that they leave a large ecological footprint have an obligation to make greater changes. As Andrew Dobson put it, "The first virtue of ecological citizenship is justice . . . a just distribution of ecological space."[43]

Justice and virtue are needed now to address climate change, as well as other environmental problems. We may not have wanted climate change, or even thought much about the impact of the industrial age and it reliance on fossil fuels. But now we know. Instead of feeling guilty or helpless, we must join with the energy industry, car manufacturers, environmental organizations, and politicians to move forward together to build a sustainable world. Echoing the immediate need for collective action, the actor Leonardo DiCaprio said, "While we are the first generation that has the technology, the scientific knowledge, and the global will to build a truly sustainable economic future for all of humanity—we are the last generation that has a chance to stop climate change before it is too late."[44]

Numerous ways exist for each one of us to help protect the earth. In the following sections, we explore three suggestions: participate in the political process, be proactive, and develop an environmental ethic. To begin requires moving from simply believing that it is important to protect the environment to being willing to do something about it. We need to be examples for others, be it in small or large ways. As Gandhi said, "As human beings, our greatness lies not so much in being able to remake the world— that is the myth of the atomic age—as in being able to remake ourselves."[45] If we change how we act, we will influence others. Each of us can embrace a humble heroism by giving the best that is in us.

Participate in the Political Process

Climate change has only recently become part of the political discourse. While elected officials at the local, state, and national levels are increasingly more vocal about the need to reduce greenhouse gas emissions, climate change and environmental protection more generally have not been a major part of most elections. During an early debate in the 2016 presidential election cycle, only one Democratic candidate, Senator Bernie Sanders, identified climate change as the most important issue facing the country. Secretary Hillary Clinton, who went on to win the nomination for the Democratic Party, mentioned nuclear weapons as her top concern. On the

Republican side, most presidential candidates refused to acknowledge that climate change exists, felt it was not due to human actions, or saw efforts to address climate change as too costly to the economy. As the Republican nominee, Donald Trump campaigned on an energy plan that would expand the use of coal and domestic oil production and relax environmental regulations. During the general election, of the debate questions posed to Hillary Clinton and Donald Trump, very few were about climate change or environmental programs.

Why do so few political candidates make environmental protection a top priority? The answer is twofold. First, the downplaying of environmental issues by presidential candidates reflects how most Americans rank the top issues facing the country. Gallup polling data consistently show that the economy remains the single leading issue in the minds of Americans. When asked in 2016 to identify the "most important problem" facing the country today, 18 percent identified the economy in general, 13 percent identified problems with the government, and 9 percent identified unemployment or jobs. Just 2 percent identified climate change or any other environmental issue as the nation's most important problem.[46] A 2016 Yale University study found that half of Americans believed that global warming will not harm them personally, and that 28 percent believed that there is a lot of scientific disagreement about global warming. Perhaps most troubling for global warming being part of the issue-attention cycle for policymakers is that fewer than 31 percent of respondents reported discussing the issue at all; 76 percent said that they heard stories about global warming in the media less than once a month.[47]

The second reason climate change is a low priority for politicians is that, sadly, many people who care about the environment and believe that climate change is a serious problem do not vote or otherwise participate in the political system. Politicians listen to voters, especially when citizens cast their ballots based on elected officials' positions on key issues. Polling data from the Environmental Voter Project show that environmentalists, as a group, are "awful voters."[48] The organization estimates that nearly sixteen million environmentalists did not vote in the 2014 midterms—a trend that was only exacerbated in the 2016 elections, which over ten million environmentalists skipped. As the organization quipped, "We can't expect environmental leadership when so few voters demand it."[49]

Thus, an essential first step as an everyday environmental hero is to vote for candidates of any party, and at all levels of government, who are committed to protecting the environment and who pay attention to how they respond to environmental issues once elected. Vote for candidates who feel the way you do about the issues you care about, and let them know when they don't. Be informed, pay attention to issues, make sure you are registered, and vote.

Although voting is the most central act of political participation, other forms are also important. Write or call your local, state, or national elected representatives, expressing your views on environmental policies and programs. In your communication, ask them what they intend to do to address climate change as well as other local, state, or regional environmental issues. Politicians pay attention when we contact them. Recent protests like the Women's March on January 21, 2017, the Environment and Science March on the forty-seventh anniversary of Earth Day in 2017, and the People's Climate Movement March on April 29, 2017, illustrate an increasing commitment of people to get involved and express their policy preferences.

As described in Chapter 6, Senator Gaylord Nelson knew the importance of putting public pressure on Congress. He understood that the massive outpouring of public support for environmental protection evidenced in the first Earth Day would be the needed catalyst to move Congress to pass tough new environmental protection laws. So it is fitting that a few years before his death, Nelson appealed to college students to be involved in the political process:

> The youth of America must be vigilant in guarding against those who would erode or erase the hard-won environmental progress in this country. . . . Assaults on the nation's environmental achievements are continuous; our defense must be unwavering. This is a special appeal to the youth of America, without whom Earth Day would not have achieved what it has achieved, and without whom the new challenge of creating a sustainable world cannot be met.[50]

Focus on Doing

Problems like climate change can seem overwhelming, causing people to doubt their ability to have any impact. However, it's important to remember the old adage that you have to move in order to go anywhere. Learning about and making sense of environmental issues from courses, books, articles, reports, research papers, and websites like NASA's is an important first step, but you must go further by sharing what you know and translating that knowledge into actions—large or small.

One way to get involved beyond voting, contacting your representatives, writing letters, and marching is to join an environmental organization. Like Judy Bonds, you can join an organization that fights for the issues you care about, adding your voice to strengthen the chorus of voices. Environmental organizations vary widely in membership size, focus, type, and locality. For example, the oldest and largest environmental group in America is the Sierra Club, with over two million members and supporters and sixty-four local chapters nationwide. Founded in 1892 by John Muir, one of America's most influential conservationists, the Sierra Club works on reducing fossil fuel use, stopping toxic chemical pollution, protecting wild places, and more. The biggest environmental organizations in the country, besides the Sierra Club, are the Defenders of Wildlife, the Environmental Defense Fund, Greenpeace, the National Audubon Society, the National Wildlife Federation, the Natural Resources Defense Council, the Nature Conservancy, the Wilderness Society, and the World Wildlife Fund (WWF)—known collectively as "The Big Green" or the "Group of Ten." WWF is the largest global environmental organization, with over five million members. This group seeks to protect species and their habitats, advance conservation around the world, and empower local communities.

These groups represent just one place to start. GuideStar, an information resource on nonprofit organizations in the United States, lists over twenty thousand conservation and environmental organizations in its directory. Choose one, or look for one in your community. Local or regionally focused environmental groups often offer the best opportunity to get involved.

Many colleges and universities have student-led environmental groups, which can have a great impact locally as well as nationally. Environmental groups on college campuses have pressed their institutions of higher learning to become more sustainable in many ways. One recent example is college students organizing to get their school's endowment funds to divest from fossil fuel stocks. As of 2013, 256 college campuses had active divestment fights under way. One goal of these divestment efforts is to put fossil fuel companies on notice that students are paying attention to climate change. As noted by Bill McKibben, former director of 350.org, an environmental group fighting climate change, "The fossil fuel industry may be dominant in the larger world, but on campus, it's coming up against some of its first effective opposition. Global warming has become a key topic in every discipline. . . . It's the greatest intellectual and moral problem in human history—which, if you think about it, is precisely the reason we have colleges and universities."[51]

Communities and local governments often seek volunteers for local environmental initiatives. Citizens may volunteer to conduct water quality sampling, develop Earth Day events, or build trails to encourage biking or walking instead of traveling by car. Local tree plantings, recycling days, household hazardous waste drives, paper and plastic reduction drives, and so on, are other possibilities.

Being active in an organization is worthwhile and important, but everyday environmental heroes must also change their own behavior. As mentioned previously, you can express your preferences by looking for and buying from more environmentally sensitive companies. Lifestyles Of Health And Sustainability (LOHAS) consumers value ethically produced, sustainably sourced products. Being a LOHAS consumer lets producers know that you are watching how food is produced and how the environment is protected in the process.

There are many other examples of actions you can take. For one, everyday environmental heroes conserve energy. LED lighting has great potential to save energy. By 2027, widespread use of LEDS could save about 348 terawatt-hours (TW-h) of electricity a year, or the equivalent annual electrical output of 44 large electric power plants.[52] It also conserves energy to buy fuel-efficient vehicles or improve the fuel economy of vehicles you already own by reducing the time you spend idling, making sure your tires are properly inflated, and avoiding hard accelerations. And you avoid sending harmful emissions into the air when you use public transportation, walk, or bike whenever possible. The EPA estimates that leaving your car at home just two days a week reduces your greenhouse gas emissions by an average of two tons per year.[53]

Many other simple steps can have an impact. The EPA estimates that current national recycling efforts reduce greenhouse gas emissions by 49.9 million metric tons of carbon equivalent, which is equivalent to the annual greenhouse gas emissions from 39.6 million passenger cars.[54] Recycling also reduces wastes that otherwise would be landfilled. Fixing water leaks, using low-flow toilets (about 40 percent of home water use is in flushing the toilets), turning down thermostats, reducing food waste, reducing meat consumption, bringing your own shopping bags to the store, avoiding plastic bags, bottles, and straws altogether, parking your car instead of letting it idle in a drive-up lane—all of these are low-cost yet effective ways to contribute to a cleaner environment and better world. Imagine if everyone acted in this way.

In short, believe that you can make a difference in your community, because you can. A famous quote attributed to the anthropologist Margaret Mead expresses this sentiment well: "Never doubt that a small group of committed citizens can change the world. Indeed, it's the only thing that ever has."[55] Focus on the doing. Treat every day like it is Earth Day.

Develop an Environmental Ethic

Connected with changing individual behavior and engaging in other forms of political participation is developing an environmental ethic. Ethics is a way of thinking about how to live a good life. Environmental ethics means applying our moral principles to our relationship with and behavior toward nature. As necessary as science is to understanding how the natural world works, it does not provide us with a moral compass. We choose how to treat the planet. Thus, moral thinking is needed to complement scientific fact.

Aldo Leopold stands as a prime example of a well-examined life combined with scientific study. The concluding chapter of his book *A Sand County Almanac* (1949) contains Leopold's most famous words on the relationship between humans and their environment. Arguing that "a thing is right when it tends to preserve the integrity, stability and beauty of a biotic community" and that "it is wrong when it tends otherwise," Leopold helped to define not only the conservation movement but also how we should treat all members of an ecological system. Leopold saw that a land ethic enlarges the boundaries of a community to include soils, waters, plants, and animals. Rather than simply having an economic relationship to the land, he saw that people need to develop close connections to nature. "That land is a community is the basic concept of ecology, but that land is to be loved and respected is an extension of ethics."[56]

If Aldo Leopold taught us to respect the land, Barry Commoner exhibited another kind of environmental ethic. In his book, *The Closing Circle* (1971), Commoner presented four "laws" of ecology: (1) everything is connected to everything else; (2) everything must go somewhere; (3) nature knows best; and (4) there is no such thing as a free lunch. In noting the interconnectedness in all ecosystems, the first law warns that if we damage a healthy ecosystem, we can trigger far wider problems. The second law cautions humanity to be careful about what and how we extract materials from nature, as the results may be with us forever. The third law suggests that we cannot improve on nature. Attempts to change natural systems will often be detrimental to that system. The fourth law posits that we have borrowed from nature for too long, and that our continued extraction from nature will ultimately have catastrophic consequences.

Commoner, featured on the February 2, 1970, cover of *Time* magazine under the headline "Fighting to Save Earth from Man," became one of the world's best-known ecologists.[57] Commoner argued that production for profit is the central cause of environmental degradation—pollution has deep connections to greed in a capitalist economic system. Thus, social and economic justice must be part of environmental protection efforts, including efforts like the Triple Bottom Line described earlier in the chapter.

The Norwegian philosopher Arne Naess founded the "deep ecology" movement, beginning in the 1970s. In developing a deep ecology platform, Naess argued that all living things have intrinsic value and inherent worth, and that humans have no right to reduce nature's richness and diversity except to satisfy vital needs. He distinguished deep ecology from what he saw as "shallow" environmentalism focused on controlling pollution through technology while still maintaining the economic system.[58] Followers of deep ecology believe that the present human interference with the nonhuman world is excessive and worsening, and that humans have a moral obligation to make changes.

These are but three examples of people who approached the environment from a moral point of view. Recently, the world's most prominent religious figure, Pope Francis, wrote about an environmental ethic. Pope Francis issued an encyclical on climate change in 2015, *Laudato Si'*, becoming the first pope to speak forcefully about the link between religion and global warming. He called for a transformation of polarized politics, economics, and individual lifestyles to battle climate change. Speaking of a new ecological conversion, Pope Francis wrote: "We have to realize that a true ecological approach always becomes a social approach; it must integrate questions of justice in debates on the environment, so as to hear both the cry of the earth and the cry of the poor." He also questioned those who would deny climate change, arguing that politicians lack a breadth of vision: "What would induce anyone, at this stage, to hold on to power only to be remembered for their inability to take action when it was urgent and necessary to do so?"[59] When President Trump visited the Vatican in 2017, Pope Francis gave him a copy of the encyclical.

These individuals made courageous statements and then exhibited their commitment through their actions. Although each one's ethical compass was different in concept and scope, each contained the same message: protecting the environment is an ethical obligation. It is up to you to decide how you want to interact with the earth.

The Need to Work Together

Climate change requires the "rest of us" to act, and it requires us to work together. Our actions can be political as we encourage politicians to adopt policies that shift to renewable forms of energy; we can also take action by encouraging companies to shift to more sustainable approaches and by changing our individual behavior and decisions. Virtually every one of us contributes to the problem by using fossil fuel–powered cars, electricity from coal-fired power plants, and so on. The need to change to renewable sources of fuel is undeniable if we are to mitigate the effects of a warming planet.

We do encounter some obstacles when we try to encourage politicians to protect the planet. Some US politicians have long refused to act on climate change, arguing that shifting away from fossil fuels to renewables is bad for the economy. A few politicians deny global warming outright. Take Senator Jim Inhofe (R-OK), who as the previous chair of the Senate Environment and Public Works Committee sharply criticized governmental actions to reduce greenhouse gases. He famously decried man-made global warming as a hoax, penning a book in 2012 titled *The Greatest Hoax: How the Global Warming Conspiracy Threatens Your Future.*[60] President Donald Trump echoed Senator Inhofe in November 2012 by tweeting: "The concept of global warming was created by and for the Chinese in order to make US manufacturing non-competitive."[61] Other politicians have protested that they are not scientists and therefore cannot properly judge the data provided by national and international scientific organizations such as NASA, NOAA, the Intergovernmental Panel on Climate Change (IPCC), or the United Nations Environmental Program.

Affiliation with a political party is a strong predictor of how Americans view global warming. In a 2016 poll conducted by the Pew Research Center, Democrats were much more likely than Republicans to view climate change as a very serious problem (68 percent and 20 percent, respectively).[62] Democrats were more than twice as likely to believe that the effects of global warming are being felt now (53 percent to 20 percent), and nearly four times as likely to think that climate change will harm them personally (42 percent to 12 percent). Just 12 percent of Republicans polled by Pew felt that climate change would affect them, suggesting that the science of climate change has not permeated public understanding. Thus, we must find ways to address the polarized politics around the issue of climate change, working together to increase our common understanding.

In contrast, many other countries have acknowledged the impacts of climate change and do not face the same politics around the issue. In 1992, governments around the world adopted the United Nations Framework Convention on Climate Change, designed as a framework to develop international agreements to reduce greenhouse gas emissions. Subsequent global meetings—or Conferences of the Parties (COP) to the agreement—helped to bring attention to the critical nature of climate change. However, global action was slow in coming. It was not until the twenty-first meeting, COP-21, held in Paris on December 2015, that virtually every country committed to reducing its fossil fuel use. The 194 countries that signed the Paris Agreement represent 96 percent of the world's population and include the largest emitters of greenhouse gas: India, China, and the United States. These countries signaled their collective intent to limit global warming to 2 degrees Celsius, while acknowledging that 1.5 degrees was the goal. The agreement went into force on November 4, 2016, with 123 parties to the agreement (including the United States) ratifying it in their country, signifying their intent to be legally bound to the terms of the treaty.

The outcome of the Paris Agreement, however, is not certain, despite its ratification. As of 2017, it remains to be seen if sufficient political will exists in the United States to fulfill the Paris Agreement. President Trump repeatedly denounced the agreement as bad for US business and announced on June 1, 2017, that the United States would pull out of the agreement.

President Trump's announcement that he intended to pull the United States out of the climate accord was countered by a number of state and local governments, as well as businesses. More than a dozen states, thirty cities, and one hundred businesses pledged to meet the US greenhouse gas emissions targets under the Paris climate accord, despite President Trump's intention to withdraw.[63] California was among the first states to act to fill the void left at the national level. Governor Jerry Brown promised to "do everything we can to keep America on track, keep the world on track . . . do everything we can to advance our program, regardless of whatever happens in Washington."[64]

Indeed, it appears that the Trump administration's plan to pull out of the agreement had the unintended effect of drawing media and public attention to climate change. Media outlets widely reported that only three countries (Syria, Nicaragua, and now the United States) were not part of the Paris Agreement. (Nicaragua and Syria subsequently agreed to sign the agreement in late 2017.) The Trump administration's efforts to renege on the US promise to address climate change also appear to have strengthened the resolve of some state and local governments, universities, corporations, and individuals to address climate change. "It's not just the volume of actors that is increasing, it's that they are starting to coordinate in a much more integral way," noted Robert Orr, a lead climate adviser to the United Nations.[65] So, too, former New York City mayor Michael Bloomberg promised in a letter to the UN secretary-general, António Guterres, that cities, states, businesses, and civil society remain committed to driving US climate change policies.

Our shared environmental story of taking action on climate change and other environmental issues as well depends on the resolve of people around the world, whether we act through countries, state or local governments, organizations, or businesses or as individuals. It also depends on finding ways to work together and bridge political and organizational divides as we pursue our common goal of ensuring a brighter, more sustainable future.

CONCLUSION

So it's up to the "rest of us." It is time to write a new story—your story, our story. It's a story that needs to be told today and into the future. We must, as Mahatma Gandhi famously put it, be the change we wish to see in the world. The stories in this book are cautionary tales about what can happen when we treat our environment mindlessly or with willful negligence. The stories here of environmental heroes also show us how to act to protect our environment and our workplaces and what to be mindful of in our lives, our organizations, and our communities.

The famous French author Alexandre Dumas wrote, "All the world cries, 'where is the man who will save us?' . . . Don't look so far for this man; you have him at hand. This man, it is you, it is I, *it is each one of us*."[66] The world needs everyday environmental heroes; the blue planet is our only home. Every moment we are writing the pages of our environmental story; every year a new chapter begins.

Judy Bonds would frequently speak to college students, exhorting them to be active and engaged environmentalists. In fiery speeches, she would tell them: "*You* are the one you have been waiting for."[67] Expressing a similar sentiment, the environmentalist Paul Hawken observed: "Nature beckons you to be on her side. You couldn't ask for a better boss. The most unrealistic person in the world is the cynic, not the dreamer. Hope only makes sense when it doesn't make sense to be hopeful. This is your century. Take it and run as if your life depends on it."[68]

DISCUSSION QUESTIONS

1. What makes it hard for companies be highly reliable?
2. Why can climate change be referred to as a problem for the "rest of us"?
3. Which of the three actions—being a part of the political process, being proactive, and developing an environmental ethic—is most important to you?
4. Do you agree with the argument here that we need everyday environmental heroes?

5. What causes shifts in policy as a result of disasters like the four stories recounted here?

6. What is the single most important thing that can be done to prevent industrial disasters from happening in the future?

NOTES

1. The term "wicked problems" was first coined by Horst W. Rittel and Melvin M. Webber in "Dilemmas in a General Theory of Planning," *Policy Sciences* 4 (1973): 155–169.

2. Richard J. Lazarus, "Super Wicked Problems and Climate Change: Restraining the Present to Liberate the Future," *Cornell Law Review* 94, no. 5 (2009): 1153–1233.

3. Samir Shrivastava, Karan Sonpar, and Federica Pazzaglia, "Normal Accident Theory Versus High Reliability Theory: A Resolution and Call for an Open Systems View of Accidents," *Human Relations* 62, no. 9 (2009): 1357–1390, 1363, doi:10.1177/0018726709339117.

4. For more on this concept, see Deborah Vaughan, *The* Challenger *Launch Decision* (Chicago: University of Chicago Press, 1996).

5. Stuart Diamond, "The Bhopal Disaster: How It Happened," *New York Times,* January 28, 1985.

6. Charles Perrow, *Normal Accidents: Living with High-Risk Technologies* (New York: Basic Books, 1984; updated edition, Princeton, NJ: Princeton University Press, 1999), 360.

7. Ibid.

8. Abraham Lustgarten, *Run to Failure: BP and the Making of the* Deepwater Horizon *Disaster* (New York: W. W. Norton and Co., 2012), 137.

9. Ibid.

10. Bryan Walsh, "Government Report Blames BP on Oil Spill. But There's Plenty of Fault," *Time,* September 14, 2011.

11. Joanna Walters, "Welcome to Libby, Montana, the Town That Was Poisoned," *The Guardian,* March 7, 2009.

12. US Department of the Interior, "Interior Department Completes Reorganization of the Former MMS" (press release), September 30, 2011, www.doi.gov/news/pressreleases/Interior-Department-Completes-Reorganization-of-the-Former-MMS (accessed September 25, 2017).

13. US Department of the Interior, BSEE, "Reforms," www.bsee.gov/who-we-are/history/reforms (accessed February 3, 2017).

14. Josh Cable, "MSHA Has Implemented Recommendations Made After Upper Big Branch Mine Disaster," *EHS Today,* January 16, 2014, http://ehstoday.com/safety/msha-has-implemented-recommendations-made-after-upper-big-branch-mine-disaster (accessed February 3, 2017).

15. Tracie Mauriello, "Upper Big Branch Disaster Spurred Mine Safety Progress," *Pittsburgh Post-Gazette,* January 18, 2014.

16. US Senate, S. 742, 110th Cong., "Ban Asbestos in America Act of 2007," www
.govtrack.us/congress/bills/110/s742.

17. Anthony Downs, "Up and Down with Ecology: The Issue-Attention Cycle,"
Public Interest 28 (Summer 1972): 38–50.

18. Ibid., 40.

19. John Elkington, *Cannibals with Forks: The Triple Bottom Line of 21st Century
Business* (Philadelphia: New Society, 1998).

20. Benjamin Hulac, "Corporations Move to Curb Global Warming," *Scientific
American,* April 21, 2016 https://www.scientificamerican.com/article/corporations
-move-to-curb-global-warming/ (accessed November 7, 2017).

21. Hiroko Tabuchi, "With Government in Retreat, Companies Step Up on Emissions," *New York Times,* April 25, 2017.

22. David Wei, Edward Cameron, Samantha Harris, Emilie Prattico, Gareth
Scheerder, and J. Zhou, "The Paris Agreement: What It Means for Business" (New
York: We Mean Business, 2016), www.bsr.org/reports/BSR_WeMeanBusiness
_Business_Climate_Paris_Agreement_Implications.pdf.

23. "Corporate Support for a Global Agreement on Climate Change: Business
and the Paris Agreement," CDP Policy Briefing, 2015, https://b8f65cb373b1b7b15feb
-c70d8ead6ced550b4d987d7c03fcdd1d.ssl.cf3.rackcdn.com/cms/reports/documents
/000/000/826/original/corporate-support-global-agreement-on-climate-change
.pdf?1471969971.

24. Timothy F. Slaper and Tanya Hall, "The Triple Bottom Line: What Is It and
How Does It Work?" *Indiana Business Review* 86 (2011): 1.

25. Swarthmore College, Global Nonviolent Action Database, "United States
Consumers Boycott Tuna to Protect Dolphins, 1988–1990," https://nvdatabase
.swarthmore.edu/content/united-states-consumers-boycott-tuna-protect-dolphins
-1988-1990 (accessed September 25, 2017).

26. *CK* staff, "Celebrating Corporate Sustainability Leadership," *Corporate
Knights: The Magazine for Clean Capitalism* (Winter 2017), www.corporateknights
.com/reports/2017-global-100/celebrating-corporate-sustainability-leadership
-14846084/ (accessed July 30, 2017).

27. See, for example, Jeff Kauflin, "The World's Most Sustainable Companies
2017," *Forbes,* January 17, 2017.

28. Jamie Smith Hopkins, "America's Super Polluters," Center for Public Integrity, September 29, 2016, www.publicintegrity.org/2016/09/29/20248/america-s
-super-polluters (accessed July 30, 2017).

29. CNN Library, 2016. "Flint Water Crisis Fast Facts," *CNN,* June 23, 2016, www
.cnn.com/2016/03/04/us/flint-water-crisis-fast-facts/ (accessed July 21, 2016).

30. Terese Olson, "The Science Behind the Flint Water Crisis: Corrosion of Pipes,
Erosion of Trust," *Truthout,* February 4, 2016, www.truth-out.org/news/item/34704
-the-science-behind-the-flint-water-crisis-corrosion-of-pipes-erosion-of-trust (accessed July 21, 2016).

31. Paul Egan, "Flint Mayor: Cost of Lead Fix Could Hit $1.5 Billion," *Detroit Free
Press,* January 15, 2016.

32. Julia Laurie, "Meet the Mom Who Helped Expose Flint's Toxic Water Nightmare," *Mother Jones,* January 21, 2016, www.motherjones.com/politics/2016/01/mother-exposed-flint-lead-contamination-water-crisis/ (accessed July 28, 2017).

33. Laurie, "Meet the Mom Who Helped Expose Flint's Toxic Water Nightmare."

34. NASA, "NASA, NOAA Analyses Reveal Record-Shattering Global Warm Temperatures in 2015" (press release), January 20, 2016, www.nasa.gov/press-release/nasa-noaa-analyses-reveal-record-shattering-global-warm-temperatures-in-2015 (accessed June 20, 2016).

35. NOAA, National Centers for Environmental Information, "Global Climate Report: May 2016," www.ncdc.noaa.gov/sotc/global/201605 (accessed June 20, 2016).

36. Chris Arnold, "Super Hot! India Records Its Highest Temperature Ever," *NPR,* May 20, 2016, www.npr.org/sections/thetwo-way/2016/05/20/478829946/super-hot-india-records-its-highest-temperature-ever (accessed June 20, 2016).

37. NOAA, National Centers for Environmental Information, "Global Summary Information—May 2016," www.ncdc.noaa.gov/sotc/summary-info/global/201605 (accessed July 26, 2016).

38. Angela Fritz, "Harvey. Irma. Maria. Why Is This Hurricane Season So Bad?" *Washington Post,* September 23, 2017.

39. NASA, "Facts: Land Ice," Global Climate Change: Vital Signs of the Planet, https://climate.nasa.gov/vital-signs/land-ice/ (accessed June 27, 2017).

40. 350.org, "Climate Science Basics," http://350.org/about/science/ (accessed June 20, 2016).

41. Brian Kahn, "Antarctic CO_2 Hit 400 PPM for First Time in 4 Million Years," *Scientific American,* June 16, 2016 https://www.scientificamerican.com/article/antarctice-co2-hit-400-ppm-for-first-time-in-4-million-years (accessed November 7, 2017).

42. Global Footprint Network, "Footprint Basics," www.footprintnetwork.org/en/index.php/GFN/page/footprint_basics_overview/ (accessed August 2, 2016).

43. Andrew Dobson, "Ecological Citizenship," paper presented at the annual meeting of the Western Political Science Association, Portland, March 11, 2004, quoted in Parker62, "Ecological Citizenship: The Basis of a Sustainable Society," Well Sharp, September 22, 2009, https://wellsharp.wordpress.com/2009/09/22/ecological-citizenship-the-basis-of-a-sustainable-society/ (accessed August 2, 2016).

44. Lorraine Chow, "Leonardo DiCaprio: We Are the Last Generation That Has a Chance to Stop Climate Change," EcoWatch, July 21, 2016, www.ecowatch.com/leonardo-dicaprio-foundation-gala-raises-45-million-1935461575.html (accessed July 23, 2016).

45. Henrik Edberg, "Gandhi's Top 10 Fundamentals for Changing the World," The Positivity Blog, May 9, 2008, www.positivityblog.com/gandhis-top-10-fundamentals-for-changing-the-world/.

46. Justin McCarthy, "Economy Continues to Rank as Top US Problem," Gallup: Politics, May 16, 2016, www.gallup.com/poll/191513/economy-continues-rank-top-problem.aspx (accessed June 9, 2016).

47. Jennifer Marlon, Peter Howe, Matto Mildenberger, and Anthony Leiserowitz, "Yale Climate Opinion Maps—US 2016," Yale Program on Climate Change Communication, http://climatecommunication.yale.edu/visualizations-data/ycom-us -2016/ (accessed June 27, 2017).

48. Gmoke, "Environmentalists Are Awful Voters," *Daily Kos,* September 28, 2016, www.dailykos.com/story/2016/9/28/1574006/-Environmentalists-Are-Awful -Voters (accessed February 5, 2017).

49. Nathaniel Stinnett, "Why Don't Environmentalists Vote?" *WBUR Public Radio,* 2015, www.environmentalvoter.org/news/why-dont-environmentalists-vote (accessed February 5, 2017).

50. Gaylord Nelson, with Susan Campbell and Paul Wozniak, *Beyond Earth Day: Fulfilling the Promise* (Madison: University of Wisconsin Press, 2002), 159.

51. Bill McKibben, "The Case for Fossil-Fuel Divestment," *Rolling Stone,* February 22, 2013, http://rol.st/YrG707 (accessed June 28, 2017).

52. Global Stewards, "Green Eco Tips for a Healthy Planet," www.globalstewards .org/ecotips.htm (accessed June 1, 2016).

53. EPA, "What You Can Do: On the Road," updated September 29, 2016, https://19january2017snapshot.epa.gov/climatechange/what-you-can-do-road _.html.

54. EPA, "What You Can Do About Climate Change," updated September 29, 2016, https://19january2017snapshot.epa.gov/climatechange/what-you-can-do -about-climate-change_.html.

55. Nancy C. Lutkehaus, *Margaret Mead: The Making of an American Icon* (Princeton, NJ: Princeton University Press, 2008), 261.

56. Aldo Leopold Foundation, "The Land Ethic," www.aldoleopold.org/about /the-land-ethic/.

57. Simon Butler, "Barry Commoner: Scientist, Activist, Radical Ecologist," *Green Left Weekly,* October 4, 2012, www.greenleft.org.au/node/52426 (accessed August 6, 2016).

58. Alan Drengson, "Some Thought on the Deep Ecology Movement," Foundation for Deep Ecology, 2012, www.deepecology.org/deepecology.htm (accessed August 7, 2016).

59. "Encyclical Letter *Laudato Si'* of the Holy Father Francis on Care of Our Common Home," 35, 42, available at *Washington Post* staff, "Read Pope Francis's Full Document on Climate Change," *Washington Post,* June 18, 2015.

60. James Inhofe, *The Greatest Hoax: How the Global Warming Conspiracy Threatens Your Future* (Washington, DC: WND Books, 2012).

61. Louis Jacobson, "Yes, Donald Trump Did Call Climate Change a Chinese Hoax," *Politifact,* June 3, 2016, www.politifact.com/truth-o-meter/statements/2016 /jun/03/hillary-clinton/yes-donald-trump-did-call-climate-change-chinese-h/ (accessed February 5, 2017).

62. Richard Wike, "What the World Thinks About Climate Change in 7 Charts," Pew Research Center, April 18, 2016, http://pewrsr.ch/1PdpUf4 (accessed July 11, 2016).

63. Hiroko Tabuchi and Henry Fountain, "Bucking Trump, These Cities, States, and Companies Commit to Paris Accord," *New York Times,* June 1, 2017.

64. Coral Davenport and Adam Nagourney, "Fighting Trump on Climate, California Becomes a Global Force," *New York Times,* May 23, 2017.

65. Quoted in Tabuchi and Fountain, "Bucking Trump, These Cities, States, and Companies Commit to Paris Accord."

66. James Penny Boyd, *Triumphs and Wonders of the 19th Century: A Volume of Original, Historic, and Descriptive Writings Showing the Many and Marvelous Achievements Which Distinguish an Hundred Years of Material, Intellectual, Social, and Moral Progress* (Philadelphia: A. J. Holman, 1899), 685.

67. Jeff Biggers, "Thousands Pay Tribute to Judy Bonds: She Has Been to the Mountaintop—and We Must Fight Harder to Save It," *Huffington Post,* May 25, 2011, www.huffingtonpost.com/jeff-biggers/thousands-pay-tribute-to_b_804001.html (accessed June 27, 2017).

68. Paul Hawken, Commencement address delivered at Portland State University, June 13, 2009, https://shar.es/1ZnW5Y (accessed July 23, 2016).

Index